NINETEEN CLUES
GREAT TRANSFORMATIONS CAN BE ACHIEVED
THROUGH COLLECTIVE ACTION

NINETEEN CLUES

GREAT TRANSFORMATIONS CAN BE ACHIEVED THROUGH COLLECTIVE ACTION

DENNY TAYLOR

New York, NY

GARN PRESS

NEW YORK, NY

Published by Garn Press, LLC
New York, NY
www.garnpress.com

Copyright © 2014 by Garn Press, LLC

Garn Press and the Chapwoman logo are registered trademarks of Garn Press, LLC

All rights reserved. No part of this publication may be reproduced, distributed, or transmitted in any form or by any means, including photocopying, recording, or other electronic or mechanical methods, without the prior written permission of the publisher, except in the case of brief quotations embodied in critical reviews and certain other noncommercial uses permitted by copyright law. For permission requests, please send an email to Garn Press addressed "Attention: Permissions Coordinator," at *permissionscoordinator@garnpress.com*.

Book and cover design by Ben James Taylor/Garn Press
Cover art by Shelton Walsmith, "Anonymous Group No. 4", oil on canvas, private collection, California

Publisher's Cataloging-in-Publication Data
Taylor, Denny, 1947-
Nineteen clues: great transformations can be achieved through collective action / by Denny Taylor.
    p. cm.
   Includes bibliographical references.
   ISBN: 978-0-9899106-7-5 (pbk.)
   ISBN: 978-0-9899106-6-8 (e-book)
1. Climatic changes-Effect of human beings on. 2. Human Beings-Effect of climate on. 3. Children and the environment. 4. Earth scientists. 5. Teachers-Political activity. 6. Equality. I. Title.
   QC903 .T395 2014
   363.7-dc23
                 2014931076

For
Johan Rockström
"We're the first generation -- thanks to science -- to be informed that we may be undermining the stability and the ability of planet Earth to support human development as we know it. It's also good news, because the planetary risks we're facing are so large, that business as usual is not an option. In fact, we're in a phase where transformative change is necessary, which opens the window for innovation, for new ideas and new paradigms."

✽

For
Lidia Brito and Mark Stafford Smith
Co-Chairs of the *Planet Under Pressure Conference* who headed the collaborative effort to write *The State of the Planet Declaration* which concludes "Society is taking substantial risks by delaying urgent and large-scale action. We must show leadership at all levels. We must all play our parts. We urge the world to grasp this moment and make history."

✽

For
United States Congressman Sheldon Whitehouse
"I speak out on climate change each week because the cost of Congress' inaction is too high for our communities, our kids, and our futures."

✽

For
Diane Ravitch
"Never have public schools been as subject to upheaval, assault, and chaos as they are today. Unlike modern corporations, which extol creative disruption, schools need stability, not constant turnover and

change. Yet for the past dozen years, ill-advised federal and state policies have rained down on students, teachers, principals, and schools."

"No other nation in the world has inflicted so many changes or imposed so many mandates on its teachers and public schools as we have in the past dozen years. No other nation tests every student every year as we do. Our students are the most over-tested in the world."

*

And, For the
American Legislative Exchange Council
Climate Change Deniers and Public School "Privateers"
ALEC has united scientists in their struggle to inform the public about climate change, and has wakened parents and teachers to the threat to democracy of ALEC's intent to privatize public education. In response to ALEC, the United Opt Out movement and BadAss Teachers have joined with parents and the public to Save Our Schools.

*

Put Your Trust In The People.

Aren't we barbarians to burn beauty in the stove, to kill what we can't recreate? Our wit and vitality are given us to increase what there is. But what do we do? We destroy. There are less forests, waters are polluted, wildlife disappears, the climate is harsher, and each day the world is poorer and uglier. You're looking at me sarcastically. You don't believe a word I say. Well, perhaps I'm crazy. But when I pass a peasant's wood that I've saved from the axe, or hear leaves rustling in a tree that I've planted -- I feel I've helped. If in a thousand years people are happier, I'll have helped.

    Anton Chekhov

    Astrov, in *Uncle Vanya*

    Translation Jean-Claude Van Itallie

Contents

Introduction   1
Nineteen Clues   8
    First Clue   10
    Second Clue   12
    Third Clue   12
    Fourth Clue   14
    Fifth Clue   14
    Sixth Clue   17
    Seventh Clue   18
    Eighth Clue   18
    Ninth Clue   21
    Tenth Clue   24
    Eleventh Clue   30
    Twelfth Clue   33
    Thirteenth Clue   37
    Fourteenth Clue   38
    Fifteenth Clue   43
    Sixteenth Clue   47
    Seventeenth Clue   50
    Eighteenth Clue   57
    Nineteenth Clue   61
Nineteen Clues to Solving the Question of Questions (QoQ)   62
Making The Planet A Child Safe Zone   65
Earth Summit Rio+20   75
Epilogue   79
Postscript   97
References   99
Declarations on Human Rights   107

Declarations on Climate Change   108

Climate Change: Organizations and Resources   109

Resistance to the Corporate Education Revolution   111

     National Organizations and Resources   111

     National Organizations and Resources - Facebook Pages   112

     State Organizations and Resources   114

     State Organization and Resources on Facebook   117

Other Organizations and Resources   123

About The Author   124

Books by Denny Taylor   126

Introduction

Inspiring People to Action

Nineteen Clues is the seminal work that led to the establishment of Garn Press. The full title is: *Nineteen Clues: Great Transformations Can Be Achieved Through Collective Action.* The emphasis of the book is on the presentation of nineteen research based clues to *actionable knowledge* that might, quite possibly, lead to the public's participation in solving problems that the US government seems incapable to fix.

One of the grand paradoxes of human achievement that *Nineteen Clues* attempts to address is that at a time when there is a new capacity for global participation in a virtual world, human activity is jeopardizing human existence in the physical world. People are, quite literally, changing the geology of the planet, which begs the question asked by Garn:

> ***What happens when a vast and ungovernable on-line world collides with a physical world, in which human activity takes place at unprecedented and multiple geopolitical scales, involving people of widely differing, and often disconnected values, ethics, emotions, spiritual beliefs, levels of trust, interests and power?***

How do we even approach the question? At Garn Press it's the question of questions (QoQ) that confronts us every day. It's the question that both excites and yes, scares us. It keeps us on the edge, lucid and conscious that, more than at any other time in history, what happens to us and the planet will depend on what we humans do. "Actionable knowledge" we call it. That's our quest. Finding out what people can do to ensure that the planet is habitable for future generations, that the rich and powerful do not exploit them, and that the creative spirit, both on-line and off-line, continues to push the boundaries of human imagination in a constant re-Visioning of our virtual and physical worlds. You can read more about this Garn Press quest by following the link to *Inspiring People to Action* on the Garn Press website.

For many readers of eBooks the idea of reading in a multidimensional space is relatively new. It is a different kind of mindfulness, it focuses our attention on the complexity of ideas, encourages us to reject the linearity of simplistic cause-effect interpretations, and pushes us to become more comfortable with the idea of "epistemic pluralism" – which just means becoming more conscious that "ideas have many roots" and can take many *routes* along often unknown trails and pathways and can arrive in many unexpected forms.

We have entered an age in which nearly everything about us is known. Data is collected by government and corporate powerbrokers using complex network driven dynamic analytics that leave very little to chance. It is not an exaggeration to state that our minds are manipulated and very often we don't even know that it is happening. Rumors and opinions are presented as "fact", often with the overt or covert intent of changing the mindset of the public on specific issues which affect their health and well being as well as the future lives of our children and grandchildren.

It is not that the public is naive or ignorant of these complex, network-driven dynamic processes, but that the multiscale nature and the intrinsic heterogeneity (many routes, many pathways) of the underlying networks make it difficult for people to develop an intuitive understanding of these processes. Most of the time we don't know the root(s) or the route(s) information has taken, or *is* taking. We don't know what is relevant and irrelevant, and we are left trying to sift through the pulp-fiction news of celebrity pontificators, who receive huge amounts of money from news organization that have perfected the art of news as entertainment, a.k.a. propaganda, that is the anathema of evidence-based, verifiable, data driven reporting.

The fact that the public is coerced, their minds manipulated, has been freely acknowledged by political and corporate powerbrokers, as you will read in *Nineteen Clues*.

Knowing that peoples' minds are being manipulated urgently requires our engagement in the quest to find ways to *counter*-act the potential impact of propaganda on our thinking and on our ways of being. *Nineteen Clues* is a practical guide to such counter-action. It is in effect a demonstration of reading in a multidimensional space in a quest for the *roots* of the "news" using multiple *routes* to counter the propaganda that the public is exposed to every second of every day.

Eric Schmidt and Jared Cohen of Google write of this virtual world, "The Internet is the largest experiment involving anarchy in history. Hundreds of millions of people are, each minute, creating and consuming an untold amount of digital content in an online world that is not truly bound by terrestrial law".[1]

While Jaron Lanier, currently at Microsoft, bluntly asks "Given both the momentum to screw up the human world and the capability to vastly improve it, how will people behave?" Lanier writes of the choices we have to make in the "architecture of our digital networks" so that we do not "tip the balance between the opposing waves of invention and calamity".[2] Lanier leaves us in no doubt that digital technology has the potential of being used for nefarious purposes. He stresses that "Digital technology changes the way power (or an avatar of power, such as money or political office) is gained, lost, distributed, and defended in human affairs" and points out that "lately, network-empowered finance has amplified corruption and illusion".

"So," Lanier writes, "we begin with the simple question of how to design digital networks to deliver more help than harm in aligning human intention to meet great challenges."

The question challenges us to claim the right to participate in a way we could not without the connectivity of the Internet. *Nineteen Clues* is written to encourage the public's active participation, your participation, to make sure you *know* how important it is that you *know* what is happening, and to ensure that you have every opportunity to weigh the evidence and consider the consequences of human activity for people and the planet. It is about privileging knowledge that has human worth but not necessarily commercial worth, about sidelining the vested interests of politicians, business leaders, and opinion makers and going straight to the sources -- to the scientists who do the research, whose knowledge is so often obfuscated, while they themselves are denigrated because their research does not fit with neoliberal doctrine or the self-interest of the super-rich elites who are causing disequilibrium in US society by gobbling up more and more of the Earth's finite resources.

Reading Lanier, you can almost see him smiling as he keys into his text, "Digital information is really just people in disguise".

Nineteen Clues builds on this premise. Participating in the democratizing possibilities of the digital age provides us all with an opportunity to live responsibility in this historical moment, with our hearts and minds focused on the here and now, knowing full well that because of the connectivity of our endeavors, together we stand a chance of having our voices heard.

The quest for actionable knowledge begins with *you* questioning the arguments I present, and with *me* demonstrating that the *Nineteen Clues* are data driven and are based on available scientific knowledge. Before we get started here is the backstory.

During the last decade in the 20th Century and the first decade of the 21st Century, Earth system scientists around the world became increasingly concerned about the overwhelming evidence that human activity is exacerbating climate change and ecosystem destruction. At the same time these scientist also began to worry about the increasing vulnerabilities of fragile human societies, given the great acceleration of the changes that are taking place on and to the planet.

The depth of concern within the scientific community is clearly evident in the *2001 Amsterdam Declaration on Global Change*,[3] and in the *Potsdam Memorandum 2007-A Global Contract for the Great Transformation*, written by the Nobel Laureates who participated in the *Nobel Cause–Symposium Series on Global Sustainability* in Potsdam, Germany. The statement signed by the Nobel Laureates who attended the symposium sums up the urgency of the issues that the scientific community is convinced that we must address:

> Humanity is standing at a moment in history when a Great Transformation is needed to respond to the immense threat to the Earth. ... Nobel Laureates from all disciplines, high level representatives from politics and world-renowned experts have called for this transformation to begin immediately.[4]

In 2009, when the Director General of the United Nations, Ban Ki Moon, spoke at Princeton University, he said it was "a make or break year for the planet and its people". It was in 2009 that Earth system scientists from around the world had reached the same conclusion. They had reached a scientific tipping point and decided to act. And so, it was agreed that the International Council for Science (ICSU) and the International Social Science Council (ISSC) would organize a series of global initiatives that would have the express purpose of bringing to the attention of political powerbrokers and policy makers their deep concerns about the findings of their scientific research for the future of human life on Earth.

Scientists in both the physical and social sciences from around the world participated in the planning of the events that took place. The efforts of these world renowned scientists began with the unprecedented event of an on-line *Global Visioning Consultation in August, 2009*.[5] This was the precursor to the *ICSU Visioning Open Forum* at UNESCO in Paris in June, 2010.[6] I participated in both these events and have written about them in *"People and the Planet: The Great Acceleration from Adaptation to Transformation"* which will be published later this year. There were many other conferences and meetings held in 2010 and 2011, culminating in the *Planet Under Pressure* conference which took place in London in March, 2012.[7]

There was little coverage of these momentous events in the US press, and it is entirely possible that you know nothing about any of these conferences and meetings organized by this extraordinary global cadre of scientists and eminent scholars, even though the political actions they took were unprecedented in the annals of science.

Nevertheless, we pick up the story with an Op-Ed piece by Thomas Lovejoy in the *New York Times* on April 5, 2012, "The Greatest Challenge of Our Species",[8] about the 2012 *Planet Under Pressure* conference, which you will find provides the first of the nineteen clues. The conference was the culminating event organized by the global scientific community through ICSU and ISSC. A joint statement was crafted by Lidia Brito and Mark Stafford Smith at the conference that was sent to every participant for review before being finalized. It was called the *State of the Planet Declaration*,[9] and it was the final document produced before the June, 2012 *United Nations Conference on Sustainable Development, Rio+20*.[10] The hope was that world leaders would take note of the global concern of the scientific community about the fragile state of the planet and the vulnerability of humanity on Earth.

It was a moment lost in Rio, largely because of the Earth shattering US "edits" to the global agreement that was made at the conference. *Nineteen Clues* ends in Rio, but the story doesn't. In the *Epilogue*, the issues that have emerged from this reading in a multidimensional space are taken up and pursued. It is my hope that in our own peaceful way we will become part of the larger story that ends with the actions of people who recognize that this is a moment in history when great transformations can be achieved by our collective action.

Nineteen Clues

Great Transformations Can Be Achieved Through Collective Action

On April 5, 2012 the *New York Times* gave us a clue about what Americans can do to save the world. No, this is not about Paul Krugman's soothsayer economics, although he does provide some clues. No, *if* we were paying attention, we would have read the first clue as we drank our coffee, but as we were probably not paying attention we just took a sip and turned or scrolled the page. Rarely when we read do we consider the consequences of the ways we live, for example, how our overconsumption is causing catastrophic damage to the planet, or how our proclivity for doing nothing about it is bringing the entire world closer to a cataclysmic disaster.

We are on an unsustainable course, but we are not prone to consider our excesses or to ask ourselves how a society that values logic and rationality can be so blind to the imminent cascading risks that exist. We do not question our human fallibility, or ask why our human instinct for survival has not kicked in. Every synapse should be firing, but this is a crisis in which cold logic will not work. And so, worried about keeping our jobs or about finding one, we miss the first clue and rush out the door because we are late. But like a crossword puzzle, once a clue is found we look for other clues in Op-Ed pieces or editorials, or in articles by staff reporters, first in the news sources of the day, and then in on-line news archives, where information is separated into topic-generated threads. And finally, when all these information sources have been exhausted, *we can data mine the primary research.* On June 10, 2012, forty years after Watergate, Bob Woodward and Carl Bernstein appeared on CBS *Face the Nation*, and Woodward said, "We were as empirical as we could be". Forty years ago it was much more difficult, but today we have vast resources just a click of a mouse or a key away. Be proactive. Be as empirical as you can be. Go to the primary sources and get the original data. We must find out for ourselves, so that we can re-examine our reasoning, question our beliefs, and decide what action to take.

We would have had no trouble finding the first clue if we hadn't been in such a rush. Thomas Lovejoy made it easy when he wrote "The Greatest Challenge of Our Species", which was published as an Op-Ed in the *New York Times* on April 5, 2012. Lovejoy is the renowned ecologist at George Mason University who introduced us in the 1980's to the concept of "biological diversity". In the Op-Ed, Lovejoy wrote without hype about the *Planet Under Pressure* conference held in London in March, 2012. This world gathering was the science precursor to the June, 2012 *United Nations Conference on Sustainable Development Rio+20*, at which new global sustainability goals were expected to be established to take the place of the Millennium Goals,[11] which will end in 2015.[12]

In the limited number of words allowed for an Op-Ed, Lovejoy reminded us that the United States abdicated its traditional leadership position in 1997 by refusing to ratify the Kyoto Protocol that might have slowed the great acceleration of the adverse climate changes we are now experiencing.[13] He also made sure we know that the controversy over climate change in the US was a "non-issue". It was an expensive political and private PR job that cost us many years when action could have been taken. The U.N. discussions to reduce emissions began in Rio in 1992, so that's twenty years of US governmental *in*activity. Meanwhile the temperature has been rising, and at *Planet Under Pressure in 2012*, the big question was how fast and how high. Go to the website of the UK Met, and read Richard Betts. We are already experiencing the extreme weather effects. Check out the April, 2012 report of an interview of Peter Höppe, the Head of Geo Risks Research for the Munich Re insurance company.[14] Höppe stated that in 2011, a regional comparison on a global scale showed that the largest increase in weather-related catastrophes occurred here, in the US. Remember the tornado that struck Joplin, Missouri in May, 2011.

At the *Planet Under Pressure* conference scientists were clear. We have left the 10,000 year epoch of the Holocene[15] and we have entered the human-driven epoch of the Anthropocene.[16] Humanity has reached the endgame unless there is immediate action. It is imperative that we follow up. Skip the blogs and go to the sources, starting with the websites of IPCC, NOAA, NASA, IGBP, UK Met, WCRP, Future Earth, Diversitas, UNEP and IHDP. Lovejoy emphasized that the global crisis is graver than we imagine, and he wrote that scientists are convinced that we have already transgressed three planetary boundaries for human life on Earth.[17] In the last paragraph Lovejoy calls the planet and its people a *"Biophysical system"*, which is a well-supported fact in Earth System Science, which leads us to our First Clue:

First Clue
People are changing the planet and the planet is changing people.

What Americans can do is to join with people of other nations around the world to sustain life on Earth. Go to the IGBP website and use the descriptors, "ecosystems", "biodiversity", and "human health and wellbeing". Research Will Steffen and Johan Rockström. The February 2012 issue of *Current Opinion in Environmental Sustainability* provides essential information[18], and the March 2012 issue of *Global Change: International Geosphere-Biosphere Programme* is of critical importance.[19]

At *Planet Under Pressure* the people-planet biophysical system was not news. Scientists focused on asking big questions like "Can science save us?" and "What can be done to sustain the planet for future generations?" The focus was on "climate extremes", "impacts of changing planetary pressures", "life in extreme environments", "disaster risk reduction" *and* "adaptation". At the conference there were intense debates taking place about "barriers to action". Delegates from all five continents participated in sessions that focused on what must be done to sustain the planet and its people. "What are the opportunities?" they asked. "What are the challenges?" Presenters spoke of the "lack of strong leadership", "deficient authority", the "lack of willingness of governments to act", "political inertia", the need for a "paradigm shift" and a "shift in discourse", "a move from national security to collective security" and a "need for global action".

In one session delegates met in small groups, and wrote on large sheets of paper about "rights and responsibilities" using descriptors including: "accountability", "cooperation", "agreements", linking "human rights" and "Earth rights. Taking turns, they wrote, "democratize", "humanise", "values versus money", and in large bold letters "stake holder participation on a planetary scale". Other delegates focused on: "living within means" with "specific goals for resource consumption". At other tables they wrote: "encourage participation", and of the importance of "bringing in individuals, not just governments". "Equity" was a recurring theme. One delegate wrote "the market has no morals". Another wrote: "capitalism and globalisation rely on rich v poor so need different system to allow equity and balance". Questions were also asked: "How can we use what we know? Combine? Make sure we are not constrained by the past?" And, "How do we measure the progress of a country, by the government or the people?"

"*We are not talking about the elephant in the room*", one delegate called out. Simultaneously, from different tables, in different parts of the room, three delegates called back, "The USA!" The atmosphere of collaboration changed to one of dissonance and dissent. There were murmurs of agreement around the room and another delegate called out, "Americans are the most overworked miserable people on the planet". Negative views of America reverberated throughout the conference, with off-the-cuff comments invariably critical. The US consumes one third of the planet's available resources, but is recalcitrant about addressing the impact of its excesses on the planet *and* its people, giving rise to anger and resentment on a global scale. People suffer because of us. Here's the Second Clue:

Second Clue
The first step to American participation in saving the world is the recognition of the legitimacy of global concerns about US overconsumption of the Earth's finite resources, and the negative impact it is having on the planet and its people.

Empirical evidence that the US is a social outlier in the developed world was presented at *Planet Under Pressure* by Richard Wilkinson. Wilkinson is gently spoken and he did not single the US out. He did not have to, because the fact that the US falls short on every social indicator that he discussed was in plain view on the graphs he used in an international comparative analysis of human well being in more equal and less equal countries. In every category the US was the most *un*equal.

Before more than 2,000 delegates, and live streamed to an audience around the world, Wilkinson made the connections between what is happening to the planet, global sustainability, and income inequality, providing us with our Third Clue:

Third Clue
Extreme social inequality in the US negatively impacts American society, increases the pressures on the planet, and has a cascading adverse effect on global sustainability.

Wilkinson spoke of the negative relationships between income inequality and of a series of health and well being indicators. The scientific evidence he presented has been published in peer reviewed medical and social science journals, is available on the Equality Trust website,[20] and is compiled in *The Spirit Level: Why Greater Equality Makes Societies Stronger*,[21] which Wilkinson co-authored with Kate Pickett. Again, skip the blogs and go to the source. The Equality Trust website will give you not only the original research data, but it will also provide you with citations to access the primary research studies on which the meta-analysis is based.

"We all do better in more equal societies," Wilkinson said, noting that *income inequality* is greatest in the US. He stated that mental illness is more prevalent in **un**equal countries, and that the US has the highest levels of mental illness in relation to international measures of income inequality. He reported that the use of illicit drugs is also highest in the US.

Wilkinson emphasized that **un**equal societies imprison more people than societies that are more equal, and once again the US is an outlier with more people imprisoned than any other country in the developed world. Violent crime is higher in unequal societies, and the US is number one in homicides per million, with a child killed by a gun every three hours in the US in 2005-2006. Obesity is higher in unequal societies, and again the US has the highest rates of obesity in the developed world. Teenage births are higher in unequal states, and again the US stands alone with by far the highest teenage pregnancy rates.

In Wilkinson's presentation he stated that child well-being is better in more equal societies, and from the graphics he presented it was clearly evident that children in the in the US live in highly stressful environments that negatively impact their everyday lives. "Improvement in child well-being in rich societies will depend more on reduction in inequality than on further economic growth," Wilkinson states on the Equality Trust website. When combined with the three previous clues, there you will also find the basis for our Fourth Clue:

> **Fourth Clue**
> **In the US further economic growth will not improve the health or well-being of the American people. Greater emphasis on income equality and less emphasis on economic growth will diminish US over-exploitation of planet's irreplaceable resources.**

Wilkinson spoke of *social mobility*, which is greater in more equal societies. Again, the US does not fare well. He also presented the evidence for education[22] that "Children do better at school in more equal societies". "Disadvantaged children do less well at school and miss out on the benefits of education," Wilkinson states on the website. "In an international analysis published in *Lancet*,[23] and an analysis of the 50 US states published in *Social Science and Medicine*,[24] we have shown that scores in maths and reading are related to inequality".

Other measures Wilkinson presented include *cohesion and trust*, with the US society one of the most fragmented and distrustful. "There has been little recognition that greater equality is an important pre-condition for strengthening community life," the Equality Trust website states, where the interconnectedness of the measures is made evident. "High levels of trust are linked to low levels of inequality, both internationally and among the 50 US states, and trust is linked to health and well-being." And so, our Fifth Clue:

> **Fifth Clue**
> **Inequality in America is bad for the planet as well as for people, and increases our ethical responsibility to act.**

Wilkinson closed the circle on human well being by emphasizing the urgent need to reduce the human pressures on the planet. He argued that for a better quality of life, we need greater income equality. Once again the Equality Trust website provides an accurate reiteration of the data that he presented at *Planet Under Pressure* to support the finding that measures of well-being or of happiness no longer rise with economic growth:

> Not only has economic growth in the rich countries ceased to bring the social benefits it once brought (and continues to bring in poorer countries), but it now threatens the planet. We are therefore the first generation to have to find ways of improving the real quality of life. The evidence suggests that we need to shift our attention away from increasing material wealth, to the social environment and the quality of social relations in our societies. For rich countries to get even richer makes little or no difference to the prevalence of health and social problems but, as other pages on this web site make clear, the social problems which beset many rich societies are much more common in more unequal societies. …
>
> It is sometimes said that societies have to choose between greater equality and economic growth. If that were true, people in the rich countries have clearly reached a point where the rational choice would be equality: If our aim is to improve the quality of life while avoiding further damage to the planet, greater equality can do both whereas economic growth can do neither.

Deftly, and without fanfare, Wilkinson provides the answer to how Americans can save the world, not alone, but in cooperation with other countries, in a global effort to slow the changes to the planet that will make it difficult for future generations to inhabit. If only it were that simple.

Turn the pages of the *New York Times* to May 17, 2010, "Motherhood: Norway Tops List of the Best Places to be a Mother; Afghanistan Rates Worst"[25] in which Donald G. McNeil Jr. gives a brief synopsis of the Save the Children annual *State of the World's Mothers* 2010 report.[26] Out of 160 countries, the US ranks below nearly all of Western Europe and just below most of the former Soviet bloc countries. America does not take care of its own. Even Cuba, after more than fifty years of isolation and US embargoes, outranked us. McNeil writes that "after Norway and Australia, the top-rated countries were Iceland, Sweden, Denmark, New Zealand, Finland and the Netherlands".

If we are proactive, go right to the source. The *State of the World's Mothers* 2012 report[27] makes the case that neither women's rights nor children's rights are protected in the US. The report states, "Apart from the United States, all developed countries now have laws mandating some form of paid compensation for women after giving birth". The report points out that the US "has the least generous maternity leave policy of any wealthy nation" and is "one of only a handful of countries in the world that does not guarantee working mothers paid leave" (p. 51).

The US is ranked last or "poor" on nearly every measure. The *lifetime risk of maternal mortality is higher* in the US than any other industrialized nation. The report states, "A woman in the US is more than 7 times as likely as a woman in Ireland or Italy to die from a pregnancy-related cause and her risk of maternal death is 15 times that of a woman in Greece". The report also focuses on children under five. More children die before the age of five in the US than in nearly every other developed country. Forty countries performed better than the US, and on this indicator of childhood mortality in the US is on par with Bosnia and Herzegovina.

The *State of the World's Mothers* report brings home our Sixth Clue:

Sixth Clue
While poor women and children in the US are the most disadvantaged, all American women and children are disadvantaged when compared with women and children in other developed countries.

In the US Mothers have the least amount of maternity leave and no guarantee of wage benefits. The estimated female to male ratio of earned income is amongst the lowest, and the percentage of women in national government is well below that of most countries in the developed world.

Digging deeper, *the US is the only country in Western Hemisphere* and the only industrialized "democracy" that *has not ratified* the December, 1979 U.N. Convention on the Elimination of All Forms of Discrimination Against Women (CEDAW),[28] and *has not ratified* the November, 1989 U.N. Convention on the Rights of the Child.[29] One argument made in the US by right wing adversaries to the ratification of these documents of basic human rights is that both women and children are protected by federal and state laws, but Wilkinson provides irrefutable evidence that they are not. When will human rights become a woman's right? When will the US ratify the U.N. Treaty for the Rights of Women in the World, and no longer stand alongside Iran and Sudan as one of only seven countries in the world that have refused to sign the treaty? When will human rights become a child's right? When will the US stand down as the only country in the world that has not signed the U.N. Convention on the Rights of the Child?

When we combine the *New York Times* April 6, 2012 Op-Ed, "The Greatest Challenge of Our Species", with the May 17, 2010 Op-Ed on the *State of the World's Mothers* report, and with the sleuthing for clues on the primary source websites, we know what we must do and we have our Seventh Clue:

Seventh Clue
The planet cannot be protected unless the rights of women and children are protected.

Let's go back to the *New York Times* and an editorial on May 20, 2012 titled "The Attack is Real: The Republican assault on women's rights and health is undeniable, severe and continuing", but which now in the online version is dated May 19, 2012, and is titled "The Campaign Against Women".[30] The editorial is unequivocal: "New laws in some states could mean a death sentence for a woman who suffers a life threatening condition". The editorial also focuses on women's lack of access to health care, their unequal pay, and their lack of protection from domestic violence. The editorial ends as follows: "Whether this pattern of disturbing developments constitutes a war on women is a political argument. That women's rights and health are casualties of Republican policy is indisputable". The hostile ideological stance of Republicans on women's rights is similar to the pattern of the Republican attack on Earth system scientists over climate change. Given this, it is not much of a stretch to consider that the elephant in the room at the *Planet Under Pressure* conference was the Republican Party, which has delayed action on climate change and global warming, and has increased the gap between the health and well being of women and children in the US and women and children in the rest of the developed world. So here is our Eighth Clue:

Eighth Clue
In the US the war on climate science and on the health and wellbeing of women and children is a symptom of a pathological political ideology that negatively impacts global stability and sustainability of life on the planet.

This clue complicates our search for other clues. Cold logic will not work. In the past, men of power who we think of as rational have come close to the total destruction of their own societies. Robert McNamara made this statement in *The Fog of War* in his recounting of the nuclear Cuban Missile Crisis.[31] He also stated that the same danger exists today. The threat of nuclear war has not gone away, but it is entirely possible that the human assault on the planet, on land, sea, and air, might be even more devastating.

Which brings us not to a clue, but to a variation of the enduring question that first appeared in the ICSU/ISSC Grand Challenges:[32]

How can timely actions be undertaken at unprecedented and multiple geopolitical scales, when the issues involve people of widely differing-and-disconnected values, ethics, emotions, spiritual beliefs, levels of trust, interests and power?

It is the Question of Questions, the *QoQ*, the unwritten question that dogged McNamara, and that President Kennedy responded to even though he did not have an answer, on a day that could have ended in nuclear annihilation. "We looked down the gun barrel to nuclear war," McNamara said in *The Fog of War*. "I want to say, and this is very important," he said. "At the end we lucked out. It was luck that prevented nuclear war. We came that close to nuclear war at the end."

In May 2012, the American Association for the Advancement of Science (AAAS) reported that the US House of Representatives had approved Energy and Water[33] and Homeland Security[34] spending bills. Paraphrasing, the AAAS estimated that atomic defense-related R&D would be *increased* by 8.4 percent, while the Office of Science R&D would be *cut* by 1.6 percent. Overall R&D in the Department's energy technology programs would be *cut* by 11.6 percent. This was mostly due to funding *reductions* for renewable energy, energy efficiency, and low-carbon innovations. At the same time R&D funding would *nearly double* in FY 2012 for the Defense Nuclear Detection Office of the Department of Homeland Security. These numbers were virtually the same as those that emerged in the Senate version of the bill[35], which passed committee on May 22, 2012 and was awaiting floor action. Tedious as it might be to read, it is another indication that while the US government supports R&D for nuclear defense technology, support for R&D on renewable energy is not a high priority. A tactical error. A nuclear attack is a possibility, but the cataclysmic impact of climate change is now inevitable.

Today, when our luck is running out, the QoQ, which is ubiquitous in life, is rarely asked, certainly not in Washington where ideology blinds politicians and fogs their minds to the imminent dangers to humanity caused by the adverse anthropogenic changes to the planet. McNamara used the space between his thumb and index finger when they were almost touching to show us how close we had come to men of power totally destroying their societies and much of the world. Imagine that we are that close to an infinite number of cataclysmic events, and will be so for the next one hundred years, and the men of power are so blinded by ideology they are incapable of asking the QoQ.

The QoQ is the question that we are really trying to address when we search for clues to respond to the question: "What can people do?" We would have to find ways to move beyond the great divisions in US society. We would have to address the systemic risks to people and the planet that are the result of the multiple platforms of US powerbrokers and decision makers. We know that the President is the only man who has the power to press the nuclear button, but men of power have many buttons that they have already pressed - political, corporate, and financial - with destructive effects that have the potential to destroy human societies. Here is the Ninth Clue:

Ninth Clue
In the US the cascading effects of powerbrokers' maladaptive decision making is quite literally changing the geology of the planet, the chemistry of the air we breathe, and the water we drink.

But the decisions that are made by men of power, (and I do mean men, very few women are in the top echelon of political, corporate, and decision makers), are fraught by their own frailties and the ideological distortions of short term victories, and not the long term consequences of their actions.

It is easy to discount the cascading effects of human fallibility on the world in which we live. Many problems that are of global significance seem to be problems that we are just facing in our local communities, and it is difficult to look beyond what is happening to our families and to our children. The *QoQ* seems like an academic abstraction, but it's not. On every level, it is the question that frames the ways in which we live. So the search is on for an article in the *New York Times* that will uncover some of the negative cascading effects of button presses by political, corporate, and financial decision makers. The article that makes the case is focused on a task force that was actually co-chaired by one of America's most powerful women, but the views expressed nevertheless reflect the platform of powerful men.

Type in "Panel Says Schools' Failings Could Threaten Economy and National Security", and a link should pop up to a piece from the Associated Press which was published in the *New York Times* on March 19, 2012.[36] Here's the first two lines: "WASHINGTON (AP) — The nation's security and economic prosperity are at risk if schools do not improve, warns a report by a panel led by former Secretary of State Condoleezza Rice and Joel I. Klein, a former chancellor of New York City's school system". AP quotes from the report include: "The dominant power of the 21st century will depend on human capital", and "failure to produce that capital will undermine American security'". The takeaway from the piece is that schools are sites of *in*security and children are *human capital*. Once again, skip the blogs and go right to the March 2012 Rice-Klein report, which is titled "U.S. Education Reform and National Security".[37]

We know that human well-being and global sustainability go hand in hand, and that inequality has negative consequences, not only for people, but also for the planet, *and* that the US leads the developed world on every indicator of inequality. But the Rice-Klein Report on US K-12 public education is not about the anthropogenic challenges that the US must address, which will require a rethinking of education for *sustainable development* and *global security*. Instead, the report focuses on US *national security*, and makes no mention of the threats right now to this generation and future generations of children, as a result of the US over consumption of Earth's limited resources.

No consideration is given in the Rice-Klein report to the fact that we are transgressing planetary boundaries for human life on Earth, or that the temperature is rising and our children can expect to live on a much hotter planet than it is now, or to the fact that wars endanger ecosystems and other species as well as maim and kill people. Again, go to the source. Look up "War and Conflict" in *Sustaining Life*, by Eric Chivian and Aaron Bernstein (2008).[38] Then go to costsofwar.org and get the data on the environmental impact of war. On this site you will also find data on the financial cost of war. The US is spending more than all other countries combined on the wars we are now fighting, increasing corporate power and the opportunities for corporate profiteering, which makes the case for the perfect marriage of Klein promoting corporate profits and Rice's advocacy for the schooling of K-12 children to prepare for war.

The worldview of this report is irrationally ideological, and the rhetoric is militaristic and threatening. Children are "human capital", and the task of teachers, predominantly disenfranchised women, is to prepare them for whatever future conflict the US might have with the rest of the world. Richard Haas, President, Council on Foreign Relations, states, "this report calls on state governors, working in conjunction with the federal government, to establish a national security readiness audit that holds educators responsible for meeting national expectations in education" (p. x).

In the Chair's Preface, Rice and Klein call the report "a clarion call to the nation" (p. xiv), but their intent is not to humanize or democratize. They state, "No country in the twenty-first century can be truly secure by military might alone. The dominant power of the twenty-first century will depend on human capital. The failure to produce that capital will undermine American security" (p. xiii). It is not difficult to imagine what Wilkinson's response to this worldview would be.

The purpose is to "catalyze national change", and "mere tweaks to the status quo will not create the necessary transformation". The Rice-Klein report argues that "urgent shifts in education policy are necessary to help the country hold onto its status as an educational, economic, military, and diplomatic global leader" (pp. 5-6). Similar statements appear throughout the report. Words and phrases used in the report include "threat", "crisis, and "negative impact", not engagement, and not participation. The Task Force advises that the United States should "aggressively implement assessments that more appropriately track student outcomes" (p. 48). To ensure aggressive implementation the Task Force recommends that:

> The Defense Policy Board, which advises the secretary of defense, and other leaders from the public and private sectors should evaluate the learning standards of education in America and periodically assess

whether what and how students are learning is sufficiently rigorous to protect the country's national security interests" (p. 50).

States and schools are advised to "remain vigilant", and that "in order to catalyze reform and innovation and better safeguard American national security, it is essential to measure how well students, teachers, and schools are measuring up", and that "accountability must also engender consequences and public awareness" (p. 53).

Thus, three key recommendations are made:

1. Implement educational expectations and assessment in subjects vital to *protecting national security*;
2. Make structural changes to provide students with good choices, which the report states will enhance choice and competition; and
3. Launch a *"national security readiness audit"* to hold schools and policy makers accountable for results and to raise public awareness.

The report is nothing less than a call to arms in preparation for future wars, and not a call for arms to hold our children and protect them from the damage to the planet which we have caused. While the concept of children as human capital and K-12 public education as training for the military is deeply troubling for US parents and educators, it is also undoubtedly of serious concern to the global community, given the immediate response that is needed from the US to the great acceleration of the pressures on the planet, and the transgression of planetary boundaries for life on Earth as we know it. Which brings us to the Tenth Clue:

Tenth Clue
Global action to avoid social or planetary tipping points[39] will need to include the active participation of the US in rethinking

K-12 education to make schools more equitable and just, and to reconnect children with the natural world.

Political progress is urgently needed, but the shift in the discourse from *national security* to *global security* is unlikely to happen in the US any time soon. Washington is in no mood to participate in any collective action that might jeopardize US global supremacy. The exhaustion of twentieth century economic structures and a predilection to go to war has left the US government fragmented, stagnated, ideologically polarized, and dangerously dysfunctional.

In addition to the militarization of US public schools, the Rice-Klein report recommends privatization, competition, and market-based approaches to education reform, all of which are deeply problematic, given the global concerns about the negative impact of overconsumption on the planet. The idea that public institutions can be privately owned, that the purpose of education is to compete, or that consumerism is the basis for school reform, is not widely supported by the public. Many groups are organizing including: *Children Are More than Test Scores; Fair Test; New York Principals; Parents Across America; Save Our Schools; United Opt Out.* But the widening gap between the US government and the American people does not seem to bother those in power. The use of propaganda to manipulate public thinking is endemic in the US. The Rice-Klein report makes this case, stating that it "believes the *annual audit should be aggressively publicized* to help all members of society understand educational challenges and opportunities facing the country." The report further states:

> This *public awareness campaign should be managed by a coalition of government, business, and military leaders.* It should aim to keep everyone in the country focused on the national goal of improving

education to safeguard America's security today and in the future.

Astute use of media and communications have a proven ability to effect changes in mindsets and actions, and the group believes that a targeted, annual campaign, led by the Department of Education in collaboration with the U.S. States, the Department of Defense and State, and the intelligence agencies could have this impact (p. 55, emphasis added).

There is no doubt that while espousing democratic principles of liberty, equality, and freedom, the minds of Americans are constantly exposed to smoking gun-mushroom cloud mind manipulations by politicians, policy makers, and government agencies. Even so, the high numbers of US casualties in the armed forces, many experiencing brain injuries, limb amputation, and other crippling body conditions, combined with the deaths and casualties of military personnel in Afghanistan, makes the Task Force report on the preparation of K-12 children for armed service seem more life ending than mind bending.

"When a system rides roughshod over its own basic assumptions, supersedes its own ends, so that no remedy can be found," Jean Baudrillard (1993) writes in *The Transparency of Evil: Essays on Extreme Phenomenon*, "then we are contemplating not crisis but catastrophe".[40]

In the case of US public schools, the erosion of democratic principles by politicians and policy makers is catastrophic, but for the people of the world and the future of the planet, the cascading effects of these covert and overt mind manipulations are potentially disastrous. But the situation could be cataclysmic, when the clarion call by US policy makers and government agencies for K-12 public schools to prepare the nation's "human capital" for military service to protect US global supremacy is combined with the pressures exerted by the US government, corporations, and billionaire plutocrats to dismantle the US public school system.

But the devil is in the details, and it takes a specific example to bring to the attention of both the US and global community how far the US has slipped from its democratic principles. In a 2009 paper "Public Education Under New Management: A Typology of Educational Privatization Applied to New York City's Restructuring", Janelle Scott and Catherine DiMartino write:[41]

> Without examining the full range of privatization actors, our understanding of educational and institutional arrangements is attenuated, the shift in power relationships becomes opaque, and the profound alterations to leadership, teachers' work, and community participation in democratic governance receive insufficient attention (p.448).

Digging deeply, they describe the power structures that *un*-Earthed the New York public school system:

> In 2002, when the state legislature gave the mayor control of the public schools, he became the ultimate gatekeeper in New York City. Upon gaining control of the public schools, Mayor Bloomberg, the former CEO and founder of Bloomberg LLP, chose to hire corporate sector professionals to be key leaders within the Department of Education (DOE). For example, he hired Joel Klein, the chairman and CEO of Bertelsmann, Inc., to be chancellor of the New York City Public Schools. In turn, Chancellor Klein hired McKinsey and Company, and Alvarez and Marshal, private management consulting firms, to help with the reorganization of governance and operational structures within the NYC DOE. Chancellor Klein hired Chris Cerf, the former president of the EMO, Edison Schools, Inc., to be the deputy chancellor of operational strategy, human capital and external

affairs. Espousing market ideologies and the positive potential of competition, these leaders invited private sector organizations to partner with the DOE to provide educational services to further their vision of schools reform (p.441).

Scott and DiMartino provide the private sector, market driven, ideological connection between Klein the CEO, Klein the Chancellor of NYC Public Schools, and Klein the Co-Chair of the US Education Reform and National Security Task Force Report. Add the admitted mind manipulations to militarize K-12 schools, and all that stands between US democracy and plutocracy are the teachers and parents of America's school children, who are doing their best to resist.

Back to the *New York Times* and the May 7, 2012 article, "Steering Murdoch in Scandal, Klein Put School Goals Aside", by Amy Chozick.[42] "While Mr. Klein still worked for Mayor Michael R. Bloomberg, Mr. Murdoch and Mr. Klein became close friends," Chozick writes. "They talked frequently about the state of public schools and Mr. Klein was lured to New Corporation with the promise that he could use the company's deep coffers to put in place his vision of revolutionizing K-12 education. Mr. Murdoch said he would be "thrilled" if education were to account for 10 percent of News Corporation's $34 billion annual revenue in the next five years". Klein was paid more than $4.5 million by Murdoch in 2011, and so a principal advisor to Murdoch, who the British House of Commons parliamentary report[43] concluded was "not a fit person to exercise the stewardship of a major international company" (Section 4, paragraph 229), continues to have enormous influence on the US K-12 public education system.

Given that the News Corp scandal began with the hacking of the phone of 13 year old Milly Dowler who had been brutally murdered, there is something perverse about the fact that Wireless Generation, for which Murdoch's News Corporation paid $360 million, has Klein at the helm in the development and use of educational data systems and assessment tools used in US K-12 public schools. Chozick writes, "Mr. Klein's education unit is now one of the few areas within the company that is currently growing, both through investment in Wireless Generation and potential acquisitions". She also states that "Wireless generation said more than 2,500 United States school districts, 200,000 teachers and three million schoolchildren currently use its products".

Back again to the *New York Times*, and a May 11, 2012 article, "E-Mails Provide Inside Look at Mayor's Charter School Battle", by Anna M. Phillips.[44] The e-mails were written during Bloomberg's 2010 campaign to expand charter schools, and Phillips writes that they were obtained through the Freedom of Information Act. The "fight of our life", Phillips writes, was the way one email described it. "We need to mobilize," Phillips reports that Klein wrote to James Merriman, the head of the New York City Charter School Center, on January 18, 2010, "Every time we keep our powder dry, we shoot ourselves." The following dialogue ends Phillip's article:

> "You were terrific," Mr. Klein wrote to Bradley Tusk, a consultant for Education Reform Now. "Perfect pitch, perfect message."

> "Who's the heavy breather on the call?" wrote a participant, whose name was redacted. "Normally, I'd ask them to mute their phone but I don't want to alienate any donors."

> "Some overweight billionaire," Mr. Klein replied.

Go to the source. The emails provide concrete verification of the nefarious activities of political, corporate, and financial powerbrokers, and the gendering of the struggle that is taking place in the US for the health and well being of American children as well as their academic development. The email exchange is grossly disparaging of the women scholars and educators who are vocal in protecting the rights of children.

Public schools by definition belong to the people, and cannot be owned by the private sector. The extremely rich cannot own the extremely poor, nor can they use the poor to increase corporate profits. In Wilkinson's international research on equality the US is an outlier, but that descriptor does not come close to describing the scorched earth policies of the business elites, working with billionaires and the Federal and state governments, destroying any hope that children in US public schools might have of responding to the challenges they will face on a planetary scale. A hundred years of solid empirical research on child development has been trashed, and extensive peer-reviewed research on human learning in the fields of anthropology, linguistics, medicine, psychology, and sociology has been thrown out.

If people in America do only one thing to save the planet, they should protect both their children and their teachers in public schools, because it is in these public places that great transformations in human learning and understanding can take place about the relationships that exist between people and the planet. And so our Eleventh Clue:

Eleventh Clue
In teaching the young we teach ourselves, and we will come to understand that the Earth is not an infinite resource to be exploited, but a finite life force that we must care for and sustain.

Go back to Lovejoy, revisit the websites of IPCC, NOAA, NASA, IGBP, UK Met, WCRP, Future Earth, Diversitas, UNEP and IHDP, add the site for the *Planet Under Pressure* conference, read the *State of the Planet Declaration* that was crafted by Lidia Brito and Mark Stafford Smith. Follow it with the 2001 *Amsterdam Declaration on Global Change*, and then read the *Potsdam Memorandum 2007- A Global Contract for the Great Transformation*, written by the Nobel Laureates who participated in the *Nobel Cause –Symposium Series on Global Sustainability* in Potsdam, Germany. They state, "Humanity is standing at a moment in history when a Great Transformation is needed to respond to the immense threat to the Earth. ... Nobel Laureates from all disciplines, high level representatives from politics and world-renowned experts have called for this transformation to begin immediately." The Nobel Laureates state, "It is essential to remove the persisting cognitive divides ... to win over young minds ... for the well-being of the generations further down the line." These documents shed a very different light on the unenlightened Rice-Klein report's call for the militarization and privatization of K-12 public education. In a dissenting view in the Rice-Klein report, Carole Artigiani, founder of Global Kids, Inc, writes of public schools as the "bedrock" of communities in "an interconnected, global society". She states:

> The current political environment is a clear demonstration of what happens when we have a public - and public officials - who are uniformed and/or ill-informed about our nation's history, our political system, and the values upon which it was built.
>
> Certainly schools must play a critical role in assuring that these needs of national security can be met. Yet, while some of the data are disturbing, nothing in this report convinces me that our public schools "constitute a very grave national security threat facing this country." Indeed, claims of alarm can only set the state for

dramatic actions unsupported by evidence: in this case, market-based approaches to school reform, that, overall, have not demonstrated their effectiveness. Indeed, charter schools and vouchers are diverting funds and energy away from neighborhood schools, and the more successful ones rely on additional support from private sources ("voluntary taxation"), a situation that is neither sustainable nor scalable. Moreover, the drive toward "competition" can diminish individual commitment to the common good, thus undermining the very nature and purpose of public education: preparing young people of all background to become informed and active citizens who understand their rights and responsibilities to contribute to society and participate in the shaping of policies that affect their communities and larger world (pp.60-61).

Stephen Walt, Harvard Kennedy School; Linda Darling-Hammond, Stanford University; and Randi Weingarten, American Federation of Teachers; all signed Artigiani's dissenting view. In his own dissenting statement, Walt writes that the recommendation of privatization encourages "a policy course that could do more harm than good" and describes the recommendations as "not a reliable blueprint for reform" (p. 66). And Weingarten writes, "In this country, no other public service essential to the nation's well-being—not law enforcement, firefighting, or armed forces—has forsaken being a public entity. Public education has been a cornerstone of democracy and a means of acculturation for generations of Americans, as well as a crucial vehicle by which those generations have not simply dreamed their dreams but achieved them (p. 69).

In her dissenting view, Linda Darling-Hammond first notes her agreement with the Task force on the importance of science, technology, engineering, and foreign languages as well as the English language arts and mathematics, before focusing on her opposition to public school privatization. "The report ignores the fact that our highest-achieving states have all built high-quality systems without charters, vouchers, educational management companies, or other forms of privatization", she states. "The path forward should be focused on building capacity to ensure high-quality options in all schools within a robust public education sector, as all high-achieving nations have done" (p. 63).

A vivid portrait of inequality in the US that is damaging people and the planet is provided by Darling-Hammond in "Why is Congress Redlining Our Schools?" published in *The Nation* on January 30, 2012, where she writes of racial and economic segregation.[45] Here, two quotes from the article underscore the urgency of the actions that need to be taken and when combined form our Twelfth Clue:

Twelfth Clue
One in four American children lives in poverty, nearly 60 percent more than in 1974, and the number of people living in severe poverty has reached a record high. A national study in 2009 found that one in fifty children in America is homeless and living in a shelter, motel, car, shared housing, abandoned building, park or orphanage. The proportions in some school districts exceed one in ten, and the number is growing rapidly.

Darling-Hammond writes of poverty concentrated in increasingly re-segregated communities, with more than 70 percent of black and Latino students attending predominantly minority schools. Writing of California, she states, "With inadequate budgets, crumbling buildings, class sizes of more than thirty (in some cases nearing fifty) and not enough desks or books, many schools serving the neediest students have long ago canceled art, music and physical education, shut down libraries and fired librarians, nurses and counselors". Inexorably, she continues writing of California, but it could be New York, any state, any city, small town or rural community, "They have lost reading specialists, science teachers, and school psychologists. As they suffer cut after cut while they seek to meet the needs of children who are often hungry and homeless as well as shortchanged in terms of educational opportunities, these schools must decide *how* they will underserve their students, not *whether* they will".

Writing of test-and-punish school reforms, Darling-Hammond states, "Blaming teachers for the ills of high-need schools lets policy-makers off the hook and keeps the more fundamental problems of severe poverty, a tattered safety net and inequitable funding under the rug". Poverty is ubiquitous, and the myths perpetuated by "blame the victim" policies, practices, and reforms are endemic in the US. The country is a sick place for children to have to go to school. There is credible evidence that in Texas, students who test low have been ousted from their schools, but what happened to the New Orleans public schools in the aftermath of Hurricane Katrina takes the US to a new moral low. Darling-Hammond writes, "In the brave new world of New Orleans, composed almost entirely of charter schools, the Southern Poverty Law Center had to sue because disabled students could not get access to public education".

In countries that care more about equality, the political status of women is more equitable with that of men, and the health and well being of children, as well as their academic development, is a priority. But in the US, where women's lives are increasingly placed in jeopardy, teachers, *who are predominantly women*, are vilified, denigrated, and bullied by political and corporate powerbrokers. Many teachers are being "pink slipped" or "excessed". In K-12 "public" schools, children are beaten up by batteries of for-profit, empirically indefensible, for-profit commercial tests, mandated by Federal and state governments, that provide *no* useful pedagogical information that can be used by teachers as a basis for instruction. Increasingly, administrators and teachers are expressing alarm at the visible impact on K-12 students, many of whom are suffering from test taking anxiety disorders. Public schools have become 21st century sweat shops in the mass production of lucrative fill in the bubble test items that distort the learning process. Again, go to the source. Click on or type in the "The Hare and the Pineapple", which is an exemplar of this lucrative sweat shop activity.[46]

The trail of pedagogical malpractice can be traced back to the 1990's, when George W. Bush was Governor of Texas. With the White House in his future, education became the center piece of his political agenda. With the help of McGraw-Hill – the Bushes and McGraws were old family friends – and with the participation of Reid Lyon, then head of the Child Development and Behavior Branch at NICHD, research was conducted in Texas on how young children learn to read. The fudging of data is documented in *Beginning to Read and the Spin Doctors of Science*.[47] The media campaign made sure that the findings supported the use of commercial skill and drill programs, and it privileged *Open Court*, which was published by McGraw-Hill. With billions of dollars at stake the PR was relentless, and the big text book adoption states - Texas, California, Florida, Michigan and New York - quickly fell. The genesis of current nefarious practices in the commercialization and privatization of the US K-12 public school system can be traced back to that time.

Despite protests from educators, the 1998 *Reading Excellence Act*, and the *No Child Left Behind Act of 2001*, which was pushed into law by the Bush Administration and the Republican controlled Congress, were based on the falsified evidence of the Texas reading studies. The *American Recovery and Reinvestment Act of 2009 (ARRA)*, more commonly called the *Race to the Top Act*, is built on the same findings, and today all commercial reading programs, testing programs, and teacher evaluation protocols can be traced back to the falsification of the research findings in the Texas studies. Pearson has been the biggest beneficiary of the political and corporate billions in PR used to instill and constantly reinforce beliefs about children's learning too complicated to be easily unpackaged, that make it mandatory for schools districts to purchase commercial skills programs, as well as the batteries of tests and all the evaluative paraphernalia that goes with them. Pearson produces and sells the commercial reading programs and the test preparation materials, produces and sells the tests, evaluates children's test taking, and then sells the evaluative procedures that hold teachers responsible if children fail.[48] Pearson is a key player in a multi-billion dollar, profit-producing, pedagogically indefensible enterprise, that siphons much needed funding from K-12 public schools and has cascading negative effects. The schools that are the hardest hit are the schools of children living in poverty who lack many other forms of basic social services. Because of the huge disparities in the "poor schools get less, rich schools get more" funding of public schools, critical educational funds are siphoned off into the private sector, to companies like Pearson, Media Corp, and Wireless Generation. All of which are under investigation, but the products they sell are still sold to our schools, as our schools are sold out to the private sector.

The complexity of the political-corporate-financial connections in the private takeover of public schools makes for a complex and difficult crossword puzzle, but once again it is important to follow up. One piece to read would be Gail Collins's "How Texas Inflicts Bad Textbooks on Us", in the *New York Review of Books*, June 21, 2012 for an inside view of how very rich Texans determine what children learn through text book adoptions in 50-80% of US K-12 schools.[49] Collins writes that the theory of evolution, global warming, and the separation of church and state are all regarded as "factual errors" by the Texas State School Board which is elected because of "some extremely rich Texans have gotten into the board of education election game, putting their money at the disposal of the conservative populists" (p. 18). But here, in our quest for clues in the *New York Times* I have picked up the thread in the article "Land of Cheese and Rancor: How Did Wisconsin Become the Most Politically Divisive Place in America?" by Dan Kaufman in the *New York Times Magazine*, May 27, 2012 which focuses on the activities of the American Legislative Exchange Council, known stateside in the personified form, as "ALEC".[50]

In the 1980 and 1990's, researchers in the educational community believed that the scientific evidence that they presented would counter the fudged findings of the Texas research. The relationships between politicians, the Business Round Table, and the decisions that were made about the teaching of children in US schools were difficult to grasp. Not so now, which provides us with the Thirteenth Clue:

Thirteenth Clue
The ideological doctrine of the American Legislative Exchange Council (ALEC) makes it easy to understand how the great wealth and vast power of the private sector is undermining not only the US K-12 public school system, but is also undermining the US response to climate change.

With billions at stake, hundreds of millions are going into the Republican coffers, and Mitt Romney, who was proactive in responding to climate change when he was Governor of Massachusetts, no longer thinks climate change is a threat.[51]

Kaufman, in his *New York Times Magazine* May 27, 2012 article "Land of Cheese and Rancor", reports that he interviewed Mark Pocan, who was a member of ALEC. Quoting from Kaufman's article, here's the strategy that Pocan explained that ALEC advises its members to use:

> "You have to introduce a 14-point platform," he said, "so that you can make it harder for them to focus and for the press to cover 14 different planks." He pointed to several bills introduced in the past two sessions, including one that allows more children to enroll in virtual charter schools. "It sounds good," Pocan said. "Kids could access virtual schools for home schooling. But again," he emphasized, the real purpose is "taking apart public schools, drip by drip."

Kaufman then confirms one of the key points made in this article, "Besides education, ALEC maintains seven other wide-ranging task forces, like "Tax and Fiscal Policy" and "Energy, Environment, Agriculture," which promotes, among other things, legislation opposing climate-change initiatives". Ideology has become policy. Given the increasing harmful effects on children of the corporate takeover of K-12 public schools, and the misery and deaths caused by extreme weather events resulting from climate change (as evidenced by the 2011 Munich Re Report), it would not be much of a stretch to make an argument for a Federal investigation under the RICO Act. Not likely, however. The Supreme Court's 2010 decision legalized super-PAC political campaign financing activities that had previously been considered illegal.[52]

Let's get back to the private takeover of K-12 public schools that are political sites of conflict, and which represent huge revenues and enormous profits for corporations. Following hard on the heels of ALEC is the Fourteenth Clue:

Fourteenth Clue
Equitable public education should be an inalienable right of every child that policy makers must protect. It is their responsibility not to sell children to private corporations that

force children to become captive consumers of products that damage their minds and increase the levels of anxiety.

The tests are protected, but children are not. The research evidence is solid. Between 1951 and 2002 there has been a *significant decline* in the mental health of high school students, and between 1938 and 2007 there has been a *significant decline* in the mental health of college students. [53] [54]

If children survive the ordeal of K-12 schooling and then apply to college, they fall prey to financial institutions that suck out every penny that they've got, and load them with hundreds of thousands of dollars of debt. In the US, studying at a college or university, indentures many young people to a life of servitude to banks that gouge them with high interest loans and reduce them to penury. In New York City there are college graduates living and begging on the streets. One young woman haunts me. She was standing where Broadway crosses Amsterdam. She could have been one of my graduate students, except for the city-pocked bags she was carrying, and the street grimed blanket. Each week there are more young people living on the streets. One young man's sign reads: "HOMELESS" NEED A Little HELP GETTING BACK ON MY FEET. ANYTHING YOU CAN DO TO HELP WILL BE GREATLY APPRECIATED. GOD BLESS YOU. And another: HOMELESS & HUNGRY. DOG (STEVIE) COMES FIRST. ALWAYS! EVERYTHING HELPS! THANK YOU AND GOD BLESS YOU! And another, this time of a homeless woman young enough to be my granddaughter, written in bubble letters: A LITTLE KINDNESS GOES A LONG WAY.

In the US many young people are on the streets even before they graduate. They have less debt but no skills, and no way to enter the workforce. In the country with the highest GDP in the world, we deliberately renege on our responsibility to educate our children. If students are not pushed out of school, they drop out. In 2011, 39,669 students dropped out of schools in New York City.[55] "We leave because they are getting us ready for prison not college" a high school senior from New York said when he participated in a global conversation at a nearby college. Three students presented from his school; two young women and one young man. They were dynamic speakers, passionate about social justice, articulate about civil rights violations, all hoping to go to college, even though they were receiving letters of rejection at the time that they spoke. There were about twenty high school students and after the presentation we sat and talked about their hopes and fears and their desire to go to university.

We should all feel their unease, as they wonder what will happen to them when they graduate. In the school they attend many students are homeless and hungry. They live with every indicator of inequality that Wilkinson spoke about at *Planet Under Pressure*. They get the connection between people helping people, and people helping the planet. Among their causes for social activism are petitions and letter writing to keep the neighborhood post office open, making a short film on using condoms to lower the rate of teenage pregnancy and STD's, an awareness initiative focusing on police racial profiling, and the establishment of a community garden in a part of New York City where schools are being closed in the Bloomberg-Klein public school private sector give-away.

Just before the end of the semester I visited the same high school students in New York. Their social science teacher, principal, and I worked together to recreate the pro-action global café that took place at *Planet Under Pressure*. Before the students arrived I read their "quote board", an uncensored place for them to write of their dreams and struggles:

> "When the people fear their government there is tyranny; when the government fears the people there is liberty", Thomas Jefferson/

"Stand for what you believe in, even if your stand alone", unknown.

"Knowledge speaks; but wisdom listens", Jimi Hendrix.

"I don't let school get in the way of my education", Mark Twain.

"Sometimes when you have too much the good gets lost within you. But when you have a little, then the good does not have to be looked for so hard", DTMS (Does That Make Sense).

"Put politicians on minimum wage and watch how fast they change", unknown.

"Why tiptoe through life to arrive at death safely?"

We created the global café with two junior classes and a senior class also joined us. We talked about the research of Earth system scientists and about Wilkinson's presentation on the negative impact of inequality on the planet. Then the students focused on the questions asked at the global café at *Planet Under Pressure*. What are the issues that are most important to your lives? What are the opportunities? What are the challenges? What can we achieve together that we can't achieve alone? In the crowded room the students, who were from many different ethnic, racial and religious groups worked together, leaving me reflect on the QoQ and imagining the possibilities of overcoming many human problems if we create spaces for young people to have such experiences.

In a similar way to the *Planet Under Pressure* global café, each group had one large sheet of paper on which they all wrote, and a representative from each group shared the discussion that had taken place in their group during the writing. One senior began by stating his group focused on social class and that they thought every other issue they discussed was caused by the domination of one class over the other. In every presentation social class was an issue. On one piece of paper, a student had written: "class warfare caused by ignorant closed minded people", another had written, "when the rich rob from the poor its called business, when the poor fight back its called violence". Students across the groups wrote and talked about: "racial profiling"; "police brutality"; "getting harassed"; "racism makes you suspect"; "walking down the street and getting pulled over"; "the police searching us for no reason"; and "we are all equal no matter what color".

Many of the issues were personal: "I'm poor but I have the same goals as a rich person"; "overcoming fears"; "being someone you don't want to be"; "having to grow up at a young age"; "not being able to help my mother since finding a job is hard". Many concerns were expressed, but issues of class and race dominated the discussion. One young woman spoke about sexual slavery, and she said that the young women who became prostitutes in her community are no different from sexual slaves who are brought into the US illegally. The students also focused on what we can achieve together that we cannot achieve alone. One group wrote as a heading: "The Necessities We Strive For". They put "education" at the center of their piece of paper and connected it with an arrow to a green and blue planet which had "Safety" in red written above it. In addition to "advocacy" and "awareness" students also wrote "protest", "revolution" and "retaliation".

My lasting impression of this global café is not only of the generosity of the students and their sincerity in discussing the issues that confront them, but also of how hurt they are by the way they are positioned by US society. One young man spoke of being poor but wanting to be an intellectual. Others spoke of their anxiety about the "culture shock" they would experience when they go on to college. They expressed concern about the lack of resources in urban schools. They shared their worries about whether or not they would be as prepared for college as students from rich schools that have much greater resources. They also named one of the most affluent communities close to New York City, and expressed concern that the wealthy people who lived there do not care what happens to people who are poor.

In a May 12, 2012 letter[56] in response to a *New York Times* article[57] "Why Don't We Have Any White Kids?" Elsie (Brooklyn) writes: "The truth is that NYC doesn't care at all about anyone from the bottom class receiving an education, black, white or purple," She continues, "This city is for rich people. If you're poor, you are treated like a nuisance at best. You will be sent to schools that feel closer to prisons." She ends the letter, "The hypocrisy and denial are an abomination….NYC has become a 3rd world country and shows no signs of changing. The gap between rich and poor, black and white is absolutely staggering and deeply, deeply shameful."

Some of the high school students who participated in the global café talked about the Occupy Wall Street movement and of going to Zuccotti Park. It is their discussion of the Occupy protests that connects us to the next article in our quest for clues to what we can do to respond to what's happening to people and the planet. This time we use a news thread and travel back to 2011, to view the Occupy Wall Street movement in light of the more recent articles that have framed this work. "Protesters Against Wall Street", a *New York Times* editorial, appeared on October 8, 2011.[58] "The message - and the solutions - should be obvious to anyone who has been paying attention" the editorial states, providing the Fifteenth Clue:

Fifteenth Clue
The problem is that no one in Washington is listening.

"At this point, protest is the message: income inequality is grinding down the middle class, increasing the ranks of the poor, and threatening to create a permanent underclass, of able willing, but jobless people", the editorial states. "This initial outrage has been compounded by bailouts and by elected officials' hunger for campaign cash from Wall Street, a toxic combination that has reaffirmed the economic and political power of banks and bankers, while ordinary Americans suffer." And then, confirming Wilkinson and Pickett's empirical research, "Extreme inequality is the hallmark of a dysfunctional economy, dominated by a financial sector that is driven as much by speculation, gouging and government backing as by productive investment".

"It seems to me that the Occupy Wall Street Movement is moral in nature, that occupiers want the country to change its moral focus," George Lakoff (2011), the Berkeley cognitive scientist and linguist writes.[59] "It is easy to find useful policies; hundreds have been suggested. It is harder to find a moral focus and stick to it. If the movement is to frame itself, it should be on the basis of its moral focus, not a particular agenda or list of policy demands. If the moral focus of America changes, new people will be elected and the policies will follow. Without a change of moral focus, the conservative worldview that has brought us to the present disastrous and dangerous moment will continue to prevail".

Holding up their placards in Zuccotti Park, the protest movement brought to the attention of the American people the gross inequities in US society. "People Before Profit"; "Human Need, Not Corporate Greed"; "Exclusive Wealth and Excessive Consumption are Dying Paradigms"; "Wall Street is a Casino"; "Too Big to Fail Too Big to Allow"; "Because They Are Going to Drive Our Planet Off a Cliff"; "Our Children Deserve Better"; "We Are You"; "If Your Neighbors Are Poor, You Are Poor"; "You are going to die. What will you leave behind?" Dylan Thomas's "Do not go gentle into that good night", was written on one placard, "Rage, rage against the dying of the light".

Zuccotti Park is empty now, and on National Public Radio a reporter says that "the protesters would not dare return". But the protesters did not have to stay. Their protests are indelibly marked on the US psyche, and their refusal to accept the inequalities in US society has not gone unnoticed in other countries around the world. Human rights activists traveled from Japan to protest the human tragedy and environmental destruction caused by the Fukushima nuclear disaster, and at least one renowned Egyptian feminist scholar, Nawal El Saadawi, who is eighty years old and participated in Tahrir Square, traveled to New York to support the social justice demonstrators in Zuccotti Park. On the streets of Manhattan "revolution" is written on signs and tee shirts and "occupy" has a new meaning. In a country that has not seen mass protests since the 1960's civil rights movement and anti-Vietnam war protests, and the 1992 L.A. riot, the possibilities of speaking truth to power is in the air we breathe, and will morph, *is morphing*, in a thousand ways.

From an analysis of the placards and signs at Zuccotti Park, it is quickly evident that the Occupy Movement is living evidence of Wilkinson's *Planet Under Pressure* presentation on the importance of equality for people and the planet. The placards that protesters were holding up were authentic reflections of the inequality in the US, and provide verification that the US is an unequal outlier in the developed world. It is also quickly evident that even though few politicians still ask it, the *QoQ* is one of the questions that undergirds the Occupy Movement, uniting the protesters both with global resistance movements *and* with the Earth System science movement, to transform human societies that are damaging the possibilities for human life on Earth.

Once more, here's the *QoQ*:

How can timely actions be undertaken at unprecedented and multiple geopolitical scales, when the issues involve people of widely differing—and—disconnected values, ethics, emotions, spiritual beliefs, levels of trust, interests and power?"

It is the people-planet question that the US will not be able to address until the great disparities between the extremely wealthy and the extremely poor are addressed. The struggle for equality in the US is a global struggle that will impact what happens *on* Earth and *to* Earth for many centuries to come. It is the reason that at *Planet Under Pressure*, the US was called the "elephant in the room".

I have used articles from the *New York Times* to open up the possibilities for us to search for the primary source data that we need to make informed decisions about the profound relationships between people and the planet that make it imperative that we address the inequalities in US society that are no longer tolerable. We can argue about ideology, but when we are in a tight corner it's best to rely on science. In this case the data that has been presented to us from the physical sciences is supported by the research that has been presented to us from the social sciences. What is happening to the planet necessitates a *step change* in every aspect of our lives.

The *QoQ* has its origins in the reVisioning of the 2010 ICSU Grand Challenges that I have written about extensively in other works. The question creates many opportunities for great conversations between people to *unpackage* human enterprise, to consider how we have become so *unearthed*, why our governments vie for global power, why our financial systems have become so predatory, and why the intent and purpose of our educational systems is to ratchet up the competition between the super powers and to increase the extreme wealth of the already extremely wealthy.

But, here's the caveat. In the U. S., these conversations will not happen while billionaires and big business control the political process. The discussions will be dead on arrival if the US does not first address the issues of inequality that are crippling American society. If the people of the US do not act to redress the enormous imbalances that exist, the consequences will be devastating both nationally and internationally, for people and the planet. *But*, and here comes the Sixteenth Clue, which is much more of a declarative statement:

Sixteenth Clue
At no time in the future will US political, corporate or financial decision makers, Democrats, Republicans, or Plutocrats, be able to deny knowledge of the anthropogenic changes that are taking place, or to cover up the fact that for power, privilege, and profit they did nothing about it.

It is with considerable foreboding that I write we will all be left saying to them that they did too little, too late.

There are more clues of course, more than enough to fill the books that I am writing, but it's time to find a set of new clues, to see what we can make of both the information presented here and the sources that are referenced, to come up with a "people's course of action". In the May 24, 2012, *London Review of Books*, Malcolm Bull has written a review entitled "What is the rational response?"[60] of Stephen Gardiner's book: *A Perfect Moral Storm: The Ethical Tragedy of Climate Change*.[61] In the review, Bull asks, "Will it get warmer still? Very probably." "Is there anything we can do about it? Potentially yes." And then he writes, "Climate change skeptics are an assortment of cussed old men, mostly without relevant scientific training, who disagree with one or more of these answers" (p. 3). This is the ubiquitous idea that from the get-go disadvantages us as we contemplate collective action.

They might be cussed old men, but in the US they are formidable, with vast wealth, power, and privilege, and they have made challenging climate change a political platform. Which leaves us asking: "Given that the opponents to any action are some of the most powerful men in the history of the world, what can we do?" Bull almost shrugs and he writes, as if in answer to our question: "Even someone who both accepted anthropogenic global warming and believed that it was possible to do something about it might look at the odds and think that fatalism was the most appropriate response".

Over the years I have been criticized for presenting a problem and not following up with what can done. Once, in a conference presentation about homelessness, someone in the audience stood up and said, "By sharing their stories you have made us feel responsible, but you haven't told us what we can do". Others have made similar statements. Like the scientist at *Planet Under Pressure*, who when challenged for not focusing on what governments can do said, "The job of scientists is to provide the evidence and that's that". It has always seemed to me that suggesting possible solutions is presumptive. But this is not the time for any of us to stand back. Bull frames the problem for us when he writes, "The real question is whether such fatalism is ethically defensible". I think it is not. He ends, "Climate ethics is not morality applied but morality discovered, a new chapter in the moral education of mankind. It may tell us things we do not wish to know (about democracy, perhaps) but the future development of humanity may depend on what, if anything, it can teach us". I agree with Bull, but would add if we want to act to mitigate climate change, any discussion of ethics must begin by unpackaging our understandings of power.

It's time for a clue from Paul Krugman. In "Egos and Immorality", published in the *New York Times* on May 24, 2012,[62] Krugman calls Wall Street's elites "self-centered", "self-absorbed" and "deeply immoral". He writes, "Think about where we are right now, in the fifth year of a slump brought on by irresponsible bankers. The bankers themselves have been bailed out, but the rest of the nation continues to suffer terribly, with long-term unemployment still at levels not seen since the Great Depression, with a whole cohort of young Americans graduating into an abysmal job market".

Let's return to Occupy Wall Street for a moment, Lakoff's comment on the moral focus of the movement, and to his warning that if there is not a change of moral focus, "the conservative worldview that has brought us to the present disastrous and dangerous moment will continue to prevail". What the Occupy movement teaches us is that we are less fatalistic than Bull thinks. Occupy has made us more aware that the extreme inequality of life in the US cannot be separated from the extreme inequality of wealth, the 99% and the 1%. It has also made us aware that the right to assemble is restricted, and that there are consequences to speaking truth to power, including a battering by the right wing media that grossly misrepresented and denigrated the movement and was very effective at limiting public support for it.

In *The Politics of Nonviolent Action*, Gene Sharp writes that most of us know very little about "the nature of power".[63] Sharp, who is an Oxford scholar, conducted his research under the auspices of Harvard University's Center for International Affairs. He states:

> One can see people as dependent upon the good will, the decision and the support of their government or of any other hierarchical system to which they belong. Or, conversely, one can see the government or system dependent on the people's good will, decisions and support. One can see the power of a government as emitted from the few who

stand at the pinnacle of command. Or one can see that power, in all governments, as continually rising from many parts of society. One can also see power as self perpetuating, durable, not easily or quickly controlled or destroyed. Or political power can be viewed as fragile, always dependent for its strength and existence upon a replenishment of its resources by the cooperation of a multitude of institutions and people—cooperation which may or may not continue (p.8).

Sharp supports the view that power is pluralistic, and "that political power is fragile because it depends on many groups for reinforcement of its power sources" (p. 8). In the US many of the groups that reinforce the pluralism of power are under attack. In this text I have focused on scientists and teachers because of the critical role education must play in preparing future generations for climate change, but many other groups are also under attack, including health professional and workers in union protected industries. A shift in power, a well orchestrated coup d'état from democracy to plutocracy, has taken place, and it will take the cooperation of many groups of people to reverse the shift. Here's our Seventeenth Clue:

Seventeenth Clue
There are signs of hope within US society. There are many groups of people working together and acting as our guides in re-Earthing. One site of resistance is K-12 schools. Even though the pressures are great, teachers, principals and parents are organizing, establishing websites, arranging meetings, holding rallies, signing petitions, opting out of testing, and sending unopened boxes of tests back to Pearson.

The dismantling of K-12 schools is dirty work, and "I'm scared" is the comment I have heard a lot. Nevertheless, many educators, predominantly women, are resisting the corporate takeover of public schools. Across the US many administrators and teachers are being bullied and psychologically battered, but many are standing their ground. The situation is particularly dire in New York City where teachers are "pink slipped", "excessed", schools are closed and then reopened with only some of the teachers hired back.

In a democratic society concerned about the negative impact that corporate and financial institutions are having on the planet, *public* schools cannot be private. In a society concerned about the adverse effects of corporate and financial institutions on human wellbeing, children cannot spend their days in school preparing for and taking corporate tests, which reproduce and perpetuate the pressures on the planet and the stressors on people, that are the cause of the catastrophic physical and social tipping points that we now face.

Across the US, teachers are speaking out against the harmful effects of the privatization of public schools, but they are no match for the billionaire powerbrokers, media moguls, corporate lobbyists, market strategists, and data management companies who have orchestrated the hostile takeover that is having such a negative impact on US children and youth. Because of the inter-connectivity of the issues, it is of vital importance that scientists and policy makers to work with educators to re-imagine *public* education, and to make schools *child safe zones* in which children can participate in projects and activities that will help them appreciate the intricate connections between their lives and the natural world. Public action to take back K-12 public schools would increase the pressures on policy makers to take back the power they have given away to the corporate and financial sectors.

It is not a radical position, although it will be described as one by the political, corporate and financial powerbrokers. Maybe, maybe not. In *The Fog of War*, Robert McNamara is unflinching in questioning the actions of men of power in the US. Included with the documentary is a list of ten lessons from his life in politics. Here's McNamara: "We are the most powerful nation in the world—economically, politically and militarily—and we are likely to remain so for decades ahead. But we are not omniscient". Of equal importance is McNamara's statement that, "We, the richest nation in the world, have failed in our responsibility to our poor and to the disadvantaged across the world to help them advance their welfare in the most fundamental terms of nutrition, literacy, health and employment". And finally, "corporate executives must recognize there is no contradiction between a soft heart and hard head. Of course, they have responsibilities to stockholders, but they have responsibilities to their employees, their customers and to society as a whole".

McNamara lived in tumultuous times. He spent three years in the military during the Second World War and was Secretary of Defense for seven years during the Vietnam War. He was at President Kennedy's side during the Cuban missile crisis, at Jackie Kennedy's side soon after President Kennedy was assassinated, picking the place in Arlington National Cemetery where the President was buried. He was at Lyndon Johnson's side when he inherited the Presidency and the Vietnam War. He was Secretary of Defense at a time when antiwar protesters had a national and international presence, and few would no doubt that it was the impact of the public protests across the US that played a vital role in ending the Vietnam War. The question we must now ask is: "What is the role of groups protesting today?"

In the sixth age of extinction, which is the name scientists have given to the time in which we live, are we really so fatalistic that we will stand by and witness our own demise? When do we become "dedicated to the great task remaining before us" of which Abraham Lincoln spoke in his Gettysburg Address? When will we begin to participate in "a new birth of freedom" to "ensure that government of the people, by the people, for the people, shall not perish from the earth"? When will we stand up and say that "government of the people, by the rich, for the rich" is no longer acceptable? When will we get it that we must act now if we want future generations not to perish on Earth?

In a technologically advanced society, with vast wealth and with more Nobel Laureates than any other country in the world, how can we tolerate such *in*equality? Equality is a societal responsibility. No country can be first amongst nations when it treats its people in such a callous way. In the US people go hungry even though vast amounts of food are thrown away. We cannot house them. Children live on the streets, while we build mansions like palaces, and apartments in New York City can cost more than a hundred million dollars.

Inequality reigns. I have never understood why the consequences of the huge disparities between the rich and the poor that are politically, economically, and socially constructed, do not constitute human rights violations. The US is quick to hold accountable other countries that commit human rights violations, but does not even seem to be aware of the human rights violations that take place at home. People die in the US, and they do die, not only as a consequence of a political system that denies them their human rights, but also because of a welfare system that is bureaucratically constructed to root out deviancy and fraud and not to provide basic social services to the poor. As an ethnographer I have spent much of my life working with men, women, and children who live in the margins of US society. Some were living on the streets or in abandoned buildings, others were struggling to keep their homes and feed their families. The families with whom I have worked have not had enough to eat, they have been denied medical care, and their children were not on the freedom trail, but found themselves instead in the pipeline from school to prison. Young men who could have gone to college were incarcerated, and the young man who was the most gifted among them was sent to prison for twenty years, which given the conditions, and the high recidivism, was in actuality a life sentence.

Black, white, men, women, some have died, an African American father with two children, a white mother with three children. The life history of one father who died is documented in *Growing Up Literate*,[64] and the life of a mother who died is documented in *Toxic Literacies*.[65] Here it is Laurie's struggle for life and her death that jolts us out of the abstraction of "inequality" and makes us confront the meaning of the word. Two years before I met her Laurie had cervical cancer and no health insurance. The decision was made not to operate. She was treated with external beam radiation and intracavity cesium insertion. Laurie never recovered. She was told she developed ulcerative hemorrhagic cystitis, or that the bleeding and bladder changes could be due to radiation cystitis, or to further malignancy. When I visited her in winter in subzero temperatures she was sick, wrapped in a blanket, alone with her children and no heat.

It was 1992 and Newt Gingrich and the Congressional Republican House were changing the rules. States were scrambling to reduce benefits to the poor. Laurie had to verify her eligibility for welfare and be recertified. She was asked if she had a bank account. She gave the caseworker her bank account and told him, "There was three dollars in it but that was about two years ago".

"Is there any money in it now?" the caseworker asked. "I don't know if the three dollars are still there," Laurie said.

"Do you have a burial plot or an agreement with a funeral home?"

Laurie grimaced. "No."

"What if she did?" I asked.

"Assets," the caseworker said. "A burial plot is counted as an asset." He looked straight at Laurie. "You would have to declare it."

"I don't have one," Laurie's voice was barely audible.

Several weeks later Laurie received a letter stating that she was mandated to attend GED classes at high school. She was told if she did not she would lose her welfare benefits. She enrolled in January 1993 and began attending classes, but she was in so much pain she could not sit through the three hour classes. Sometimes she was so sick she was unable to walk to school. Her case technician was notified. She got a note from her doctor but it was rejected. After several weeks of document gathering, and a letter that I wrote on her behalf, Laurie was temporarily released from her "obligation" to attend GED classes. She rarely left her home. In constant pain she lay on the couch, burning and bleeding, knowing that she would have to return to school, and that if she did not her benefits would be taken away. And so we petitioned the state to allow me to teach Laurie, although I always knew it was Laurie who was teaching me. I arranged for Laurie to receive her lessons via the postal service from a mail-order GED program, and we did them together for a while with her three children playing around us. When I left the state we kept in touch and just before *Toxic Literacies* went to press I received word that Laurie had died. There is a memorial to her at the end of the book.

I have thought of Laurie a lot in writing this book. At the *Planet Under Pressure* conference in London when Wilkinson spoke of inequality, I also thought about her. If you are poor and a woman you have no rights in America. I think of Laurie when I see the young women on the streets in New York City. They might not have cancer, but their life expectancy is foreshortened by the conditions of their lives. It is twenty years since I first worked with Laurie, and inequality has become a large gaping hole in the soul of America. "So much of what we're asked is to obey" Tracy K.

Smith wrote in her poem, *Solstice* (in *"Life on Mars"*), "Our time is brief. We dwindle by the day."[66]

In "A vision for human well-being: transition to social sustainability" in the February, 2012 issue of *Current Opinion in Environmental Sustainability* which focused on *Planet Under Pressure*, Deborah Rogers, et al. make the case that human well-being and environmental sustainability are closely connected. Cautioning us, they state:[67]

> Economic growth does not necessarily translate into expenditures that increase the well-being of members of society. Instead of spending on public health, education, infrastructure and other essential components of good lives and functioning communities, such wealth may simply be used to increase the luxuries available to a few. As resources (clean water, timber, farmland, oil reserves, and others) become less abundant, continued growth in their utilization is no longer an option. Finally, the negative impacts to the biosphere of ever-increasing production, consumption, and waste disposal, including greenhouse gases, generate significant risks to human kind as well as to biodiversity and the environment (p. 63).

Rogers and her colleagues write that "environmental sustainability requires human societies that function well". These researchers write, "We have been measuring societal success on the basis of a production indicator for more than half a century", and they reason that GDP can no longer be used to measure societal progress.

They state, "Today, there is a wide consensus in the literature that we should go beyond GDP to measure well-being in a more comprehensive way (p. 69). Finally, they advance the proposition that:

> In return for these changes, communities and societies may experience better social relationships and less conflict within and

between societies. The material demands placed on the environment can be reduced to a sustainable level. A commitment to addressing human well-being in an equitable way will make possible the kinds of joint decision-making and collaborations needed to solve the world's problems. Best of all, once success and happiness are no longer defined solely in terms of material wealth, human happiness and well-being can continue to grow without exceeding sustainability limits and planetary boundaries (p. 70).

Such great transformations in human societies will not happen without great transformations in the political, corporate, and financial power bases in the US, which are unlikely to happen anytime soon without the active encouragement of the many groups within American society that have the capacity to work together to shift the power base if they so chose.

During the global conversation at *Planet Under Pressure*, where the US was called "the elephant in the room", the delegates gave us the Eighteenth Clue:

Eighteenth Clue
The key components in the struggle for social and environmental sustainability identified by the delegates at the Planet Under Pressure global conversation were: (1) the active engagement of people; (2) the participation of diverse social groups working together; (3) a focus on human well being; (4) the development of political will; and (5) ethical and principled global agreements.

In the abstract it seems irrational to write, let alone read, partly because "adaptation" is not an intellectual construct that can be written neatly on the page and then simply carried out. It is made up of *purposeful* behaviors, thinking, willing, and *doing*. We have not faced up to the consequences to the US and to the world of the destructive behaviors of the political, corporate and financial sectors, or the corporate takeover of people's lives, exemplified here by the privatization of public schools and the indenturing of young people with debts for their education they will pay for the rest of their lives.

Twenty years ago at the 1992 Rio Earth Summit,[68] the world made a commitment to eradicate poverty (principle 5), while conserving, protecting and restoring the health and integrity of Earth's ecosystems (principle 7). The US egregiously reneged on this commitment, and now we are running out of time. It is not humanly possible to save the planet without the full cooperation and active participation of all sectors of US society.

Political progress is urgently needed, and yet in the run up to the 2012 presidential election, neither eradicating poverty nor restoring the health and integrity of ecosystems were even a footnote on the political platforms presented. We know the drivers of poverty in US society, and we can eradicate them. We know what must be done to reduce inequality, and we can make US society more equal. We know what must be done to conserve, protect and restore the integrity of Earth's ecosystems and we can do it. At both the *Planet Under Pressure* and Occupy Wall Street these were critical issues that were addressed. Making the planet **A Child Safe Zone** has become the transformative idea that gives purpose and passion to my work. If it's not good for kids it's not good for the planet. It's that simple.

In the great transformation that scientists are urging us to make, the US could begin by taking care of its own children, making sure both that they have homes to live in and food to eat, and that the government does not sell them out for corporate profit. Schools could become new sites of engagement where the stories of people and the planet are told, imparting to our children a different sort of message, a different sense of their own life histories and of their future, engaging them in projects and activities in the sciences and the arts, and rejecting the pathological obsession of grill, drill, and test them. Our children could learn to be active, critical thinkers, who ask questions about humanity and the planet, and then use their imagination and creativity to address them. For that to happen administrators, teachers, and parents will have to *reoccupy* the schools, from which they have been ousted, and dump the corporate raiders, beginning with Pearson and Wireless Generation.

We are back to the beginning, to Thomas Lovejoy and his *New York Times* April 5, 2012 Op-Ed, "The Greatest Challenge of Our Species". Lovejoy was the only American on the Board of Patrons of the *Planet Under Pressure* conference. In their endorsement of the *State of the Planet Declaration*, the Patrons state:

> The human species is degrading the environment at all spatial scales, from local to global. ... The survival of our societies, our civilizations and our cultures are dependent on a stable climate, natural resources and ecosystem services. We have become a force of nature, but individually we continue to be vulnerable. Business-as-usual is not an option. The time for action is now. ... There is no time to lose.

The *State of the Planet Declaration*, crafted by Lidia Brito and Mark Stafford Smith to represent the views of Earth System scientists from around the world, states:

> Research now demonstrates that the continued functioning of Earth system as it has supported the well-being of human civilization in recent centuries is at risk. Without urgent action, we could face threats to water, food, biodiversity and other critical resources; these threats

risk intensifying economic, ecological and social crises, creating the potential for a humanitarian emergency on a global scale (p.1).

Society is taking substantial risks by delaying urgent and large-scale action. We must show leadership at all levels. ... We urge the world to grasp this moment and make history (p 4).

It's time for one last clue. On May 9, 2012 James Hansen had an Op-Ed piece in the *New York Times* entitled, "Game Over for Climate Change".[69] Since the 1980's, Hansen has taken more criticism than any other scientist in the US because of his dogged determination to raise public awareness that human activity is imperiling the planet. "GLOBAL warming isn't a prediction. It is happening," Hansen writes. In the Op-Ed he sounds the alarm about Canada's exploitation of the oil in its vast tar sands reserves, stating without equivocation that it will mean "game over for climate change". Once again it is important to go to the primary sources. Hansen's concerns are real. He states:

> The science of the situation is clear - it's time for the politics to follow. This is a plan that can unify conservatives and liberals, environmentalists and business. Every major national science academy in the world has reported that global warming is real, caused mostly by humans, and requires urgent action. The cost of acting goes far higher the longer we wait - we can't wait any longer to avoid the worst and be judged immoral by coming generations.

In letters and on blogs he is slammed for exaggerating and derided for stating that "every major national science academy in the world" agrees that climate change is real. It is Hansen who provides the Nineteenth Clue:

Nineteenth Clue
The brute power of ideology and great wealth is used to distort and discredit science, but it cannot change the scientific evidence and is no match for the courage and endurance of scientists who stand their ground or for the people who support them.

The politically motivated media campaign to inspire and provoke the public to question climate change has been devastatingly effective. Bull is back with his "climate change skeptics are an assortment of cussed old men, mostly without relevant scientific training". They are also dangerous and destructive. Don't be taken in.

We have covered enormous ground in our search for clues to what the people of the United States can do. Much of what is written here is presented in more detail in my other writings, but I hope I have written enough to convince you to go to the primary sources and find out for yourselves the veracity of the concerns of scientists about the impact on humanity of the great acceleration of the anthropogenic changes that are taking place. In the end it is up to us. We have to ask: What can be done to sustain the planet for future generations? How can we lessen the impact of changing planetary pressures; prepare for life in extreme environments; reduce the risk of disaster, and adapt? What can we do to overcome barriers to action; political inertia; the lack of strong leadership; deficient authority; and the unwillingness of governments to act? Before addressing these questions let's take another look at the nineteen clues.

Nineteen Clues to Solving the Question of Questions (QoQ)

Here's the Question of Questions (*QoQ*): How can timely actions be undertaken at unprecedented and multiple geopolitical scales, when the issues involve people of widely differing—and—disconnected values, ethics, emotions, spiritual beliefs, levels of trust, interests and power?

FIRST CLUE: People are changing the planet and the planet is changing people.

SECOND CLUE: The first step to American participation in saving the world is the recognition of the legitimacy of global concerns about US overconsumption of the Earth's finite resources, and the negative impact it is having on the planet and its people.

THIRD CLUE: Extreme social inequality in the US negatively impacts American society, increases the pressures on the planet, and has a cascading adverse effect on global stability and sustainability.

FOURTH CLUE: In the US further economic growth will not improve the health or well-being of the American people. Greater emphasis on income equality and less emphasis on economic growth will diminish US over-exploitation of planet's irreplaceable resources.

FIFTH CLUE: Inequality in America is bad for the planet as well as for people, and increases our ethical responsibility to act.

SIXTH CLUE: While poor women and children in the US are the most disadvantaged, all American women and children are disadvantaged when compared with women and children in other developed countries.

SEVENTH CLUE: The planet cannot be protected unless the rights of women and children are protected.

EIGHTH CLUE: In the US the war on climate science and on the health and wellbeing of women and children is a symptom of a pathological political ideology that negatively impacts global stability and sustainability of life on the planet.

NINTH CLUE: In the US the cascading effects of powerbrokers' maladaptive decision making is quite literally changing the geology of the planet, the chemistry of the air we breathe, and the water we drink.

TENTH CLUE: Global action to avoid social or planetary tipping points will need to include the active participation of the US in rethinking K-12 education to make schools more equitable and just, and to reconnect children with the natural world.

ELEVENTH CLUE: In teaching the young we teach ourselves, and we will come to understand that the Earth is not an infinite resource to be exploited, but a finite life force that we must care for and sustain.

TWELFTH CLUE: One in four American children lives in poverty, nearly 60 percent more than in 1974, and the number of people living in severe poverty has reached a record high. A national study in 2009 found that one in fifty children in America is homeless and living in a shelter, motel, car, shared housing, abandoned building, park or orphanage. The proportions in some school districts exceed one in ten, and the number is growing rapidly.

THIRTEENTH CLUE: The ideological doctrine of the American Legislative Exchange Council (ALEC) makes it easy to understand how the great wealth and vast power of the private sector is undermining not only the US -12 public school system, but is also undermining the US response to climate change.

FOURTEENTH CLUE: Equitable public education should be an inalienable right of every child that policy makers must protect. It is their responsibility not to sell children to private corporations that force children to become captive consumers of products that damage their minds and increase their levels of anxiety.

FIFTEENTH CLUE: The problem is that no one in Washington is listening.

SIXTEENTH CLUE: At no time in the future will US political, corporate or financial decision makers, Democrats, Republicans, or Plutocrats, be able to deny knowledge of the anthropogenic changes that are taking place, or to cover up the fact that for power, privilege, and profit they did nothing about it.

SEVENTEENTH CLUE: There are signs of hope within US society. There are many groups of people working together and acting as our guides in re-Earthing. One site of resistance is K-12 schools. Even though the pressures are great, teachers, principals and parents are organizing, establishing websites, arranging meetings, holding rallies, signing petitions, opting out of testing, and sending unopened boxes of tests back to Pearson.

EIGHTEENTH CLUE: The key components in the struggle for social and environmental sustainability identified by the delegates at the *Planet Under Pressure* global conversation were: (1) the active engagement of people; (2) the participation of diverse social groups working together; (3) a focus on human well being; (4) the development of political will; and (5) ethical and principled global agreements.

NINETEENTH CLUE: The brute power of ideology and great wealth is used to distort and discredit science, but it cannot change the scientific evidence and is no match for the courage and endurance of scientists who stand their ground or for the people who support them.

Making The Planet A Child Safe Zone
Relearning Together How To Live In The World

What happens next is up to us, whether we opt for fatalism or activism, whether we focus on eradicating poverty, taking back our schools, or working in one of the many other areas that are of grave concern, we have to act. If we do the US will no longer be viewed as an impediment to global action, but will take its place alongside other countries, large and small, rich and poor, working for equality as Wilkinson and Pickett would have us do, not as a utopian dream, but as the best hope we have to stop the temperature rising and to sustain the wellbeing of human life on the planet. And yes, the people of the US can become part of the global community that takes seriously the idea that *if* people work together we *can* save the world.

At the *Planet Under Pressure*, Will Steffen said that there is not much time left. Steffen's consistent message is that we are facing *faster change* and *more risk*. At the conference there was no contestation of the science. Delegates agreed, as one delegate put it, that "this is serious stuff". There was agreement that we must "re-learn together", "create a new philosophy of life", "shift our paradigm", "be critical of world views and mental models", "recognize limitations of resources", and take a "seven generation view in policy making". At the global café delegates were immersed in conversation. Now, I read what they wrote as if they were shouting.

"Stop amassing unnecessary wealth, especially if you not going to share it!"

"WE NEED NON-GROWTH MODELS NOW!"

"How you believe change happens, influences the process by which you think decisions should be made?"

"Envisage the cultural values we will need in 2050 then backdate them."

"We concur on core issues so tackle inequality."

"There's a lack of equality, a lack of trust."

"So start with empowerment."

"We need to invite people to the table."

"Involve the poor in making decisions."

"Encourage people to network to facilitate participation."

"Are we communicating in ways that all can talk?"

"We need actions not words."

"Bring individuals from nations not just their governments."

"Change starts from small groups of people after reaching tipping points."

"Do you believe you can have impact?"

"Equality is absolutely needed as a prerequisite (or an outcome of a process)."

"We need to start now, holding a vision of equality."

"People need a local participatory lens, focus on what really matters to a community, and then on action and communication."

"Local progress on global indicators."

"Make it cool for popular culture, make it cool."

The clarion call at the conference was for a revolution of mind and spirit, a rethinking of our very being *in* and *of* the world, which would require a fundamental change in the ways in which we live. Almost daily now there are scientific reports and research articles published about the social and ecological disasters associated with climate change. In June, 2012 there is a press announcement from the University of Berkeley[70] and an article titled "Climate change and disruptions to global fire activity" in *Ecosphere*, an open-access, peer-reviewed journal of the Ecological Society of America, by Max Moritz, fire specialist, on the rapid increase in global fire risks that are attributable to climate change.[71] In the Berkeley press release, Sarah Yang writes "Almost all of North America and most of Europe is projected to see a jump in the frequency of wildfires, primarily because of increasing temperature trends". She quotes Moritz, who states, "In the long run, we found what most fear - increasing fire activity across large parts of the planet. But the speed and extent to which some of these changes may happen is surprising. These abrupt changes in fire patterns not only affect people's livelihoods, but they add stress to native plants and animals that are already struggling to adapt to habitat loss."

Fahrenheit 451? A dystopian twist of a dystopian novel? The nineteen clues in response to the *QoQ* challenge the misconception that the people have no power. Every social and environmental indicator points to the fact that the hostile rich-win take-over of US society is unsustainable. Stand in a book store in New York and read the titles of the new non-fiction books:

Predator Nation: Corporate Criminals, Political Corruption, and the Hijacking of America, by Charles Ferguson;

The Dictator's Learning Curve: Inside the Global Battle for Democracy, by William Dobson;

Days of Destruction, Days of Revolt, by Chris Hedges and Joe Sacco;

What Money Can't Buy: The Moral Limits of Markets, by Michael J. Sandel;

End this Depression Now!, by Paul Krugman;

The Price of Inequality How Today's Divided Society Endangers Our Future, by Josef E. Stiglitz;

The Real Crash: America's Coming Bankruptcy --- How to save Yourself and Your Country, by Peter D. Schiff;

DemoCRIPS and ReBLOODlicans: No More Gangs in Government, by Dick Russell and Jesse Ventura;

Me the People: One Man's Selfless Quest to Rewrite the Constitution of the United States of America, by Kevin Bleyer;

Do Not Ask What Good We Do: Inside the U.S. House of Representatives, by Robert Draper;

Private Empire: ExxonMobil and American Power, by Steve Coll;

As Texas Goes...How the Lone Star State Hijacked the American Agenda, by Gail Collins;

Twilight of the Elites: America after Meritocracy, by Christopher Hayes;

Cowards: What Politicians, Radicals, and the Media Refuse to Say, by Glenn Beck.

Americans have been shaken up. Great transformations in people's thinking are already taking place, but few writers are making the connections that were made between people and the planet, or between equality and social and environmental sustainability, that were made at *Planet Under Pressure*. For the sake of our children we must make social equality our life's work. It is possible that through human endeavor, equality, stability, and sustainability can become the key indicators that we use as the measure of a society's success or failure. If these indicators were used today, the US would be considered a failed state, or more troubling, a rogue state. Both descriptors apply. Aggressive competition to be first in GDP exacerbates the inequalities that destabilize not only US society, but other societies in both the developed and developing world, jeopardizing global social and environmental sustainability.

For our children and our grandchildren we have to act now to make the planet a **child safe zone**. I have called this "the impossible project". If we think of the great transformations that are needed as many, many, small changes, then we can start. One place to begin is by organizing global cafés using the frameworks which were established at the *Planet Under Pressure* and then were recreated with the junior and senior high school students in New York. Both events began with a presentation about the connections between social equality and environmental stability and sustainability. The questions that were introduced earlier in the text apply: What are the issues that are most important to your lives? What are the opportunities? What are the challenges? What can we achieve together that we can't achieve alone? It might be that an invitation is made by a grandmother who gathers her friends, and together they research the primary sources and find out as much as they can about the climate and environmental changes that are taking place, and then address the questions presented at the global café. Educators and health professional could do the same. The eighteenth clue provides a framework for what it will take if great transformations are to occur in our thinking about the connections between people and the planet. This framework will have to include: (1) the active engagement of people; (2) the participation of diverse social groups working together; (3) a focus on human well being; (4) the development of political will; and (5) ethical and principled global agreements.

The global cafés have the potential to increase the possibilities for peaceful collective action. Gene Sharp provides many examples of non-violent resistance, and the erosion of democratic principles in the US makes his texts more and more relevant to the struggle against the increasing disenfranchisement of many groups within US society. Addressing the critical issues identified both in real and virtual spaces would increase the possibility of finding ways to respond to the *QoQ*, by bringing together people with widely differing and disconnected values, ethics, emotions, spiritual beliefs, levels of trust, interests, and power who are willing to work together so that timely actions can be undertaken at unprecedented and multiple geopolitical scales.

It is entirely possible that the greatest contribution the American people can make to the reduction of the accelerating risk of the temperature rising more than 4oC (7oF) by the end of the 21st century, or to reducing the risk of the rapidly occurring transgression of planetary boundaries for life on Earth, is to reign in the US government, as well as corporations and financial institutions at home. We the people, of the people, not only for the people in the US, but for the people of the planet. What happens here matters. If the super PAC ideological elites with vast reserves of mobile capital were to win the White House, the gap between the rich and the poor will increase, regulations to keep the air and water clean will be in jeopardy, the degradation of the environment will continue at an alarming rate, and the destructive financial practices that have caused so much misery and suffering will result in the transgression of planetary boundaries.

On the front page of the *Guardian Weekly* (June 15-21, 2012) one week prior to Rio+20, the headline article by John Vidal, the *The Guardian*'s environmental reporter, read: "Ecological web is badly tangled". The same article had been previously published in *The Guardian* on June 7, 2012, called "Many treaties to save the earth, but where's the will to implement them?"[72] Vidal writes that, "ecosystem decline is increasing, climate change is speeding, soil and ocean degradation continues, air and water pollution are growing, and we are still getting sustainable development disastrously wrong". Vidal describes the United Nations Environmental Programme (UNEP) Global Environmental Outlook (GEO-5) report,[73] which is a hefty 528 page tome, but the Summary for Policymakers[74] is twenty pages and well worth the read. The UNEP June 12, 2012 press release[75] gives the gist in eight pages and states that "Little or no progress was detected" towards 24 environmental goals and objectives, "including climate change, fish stocks, desertification and drought." Vidal writes, "Governments spend years negotiating environmental agreements, then willfully ignore them. So what is the point?" He writes, "The question is, are all these agreements no more than vain promises by cynical governments to wave a piece of paper in front of gullible electorates?" Further on he writes, "Rich countries have consistently promoted a global economic agenda that deliberately opens up poor countries to powerful corporations that can lobby, bully, cajole, or just ignore national and international environmental laws and agreements". He concludes by stating "it's not in the interests of most governments to change the status quo."

Pick up the thread in "Rio+20 conference's search for green solutions hampered by deep divisions" in the *The Guardian*, June 12, 2012.[76] Jonathan Watts, *The Guardian*'s Asia environment correspondent, writes of the Rio+20's "once-in-a-generation" Earth Summit of heads of state, in which he describes the deep divisions between nations and the low expectations for success. He writes that President Obama had not confirmed that he will attend, David Cameron, the UK prime minister, will send a deputy, as will the German chancellor, Angela Merkel. Watts states that no new legally binding treaties were expected, and points out that according to UNEP, in the last two decades carbon emissions have increased 40% and biodiversity loss has risen 30%. Watts quotes the UNEP director, Achim Steiner, who warns, "If current trends continue, if current patterns of production and consumption of natural resources prevail and cannot be reversed and 'decoupled', then governments will preside over unprecedented levels of damage and degradation". Watts writes that ahead of the Earth Summit, Rio is hosting a Peoples Summit which 50,000 people from around the world were expected to attend to share best practices and make commitments to action.

In the end the human struggle for our own survival and for the survival of the planet as we know it will come down to the ways in which we think about it. In her December 7, 1993 acceptance speech for the Nobel Prize in Literature,[77] Toni Morrison writes of language withheld for "certain nefarious purposes", and she encourages us to think of "language partly as a system, partly as a living thing over which one has control, but mostly as agency - as an act with consequences". Morrison helps position us as we think about the "unyielding" language of political, corporate, and financial decision makers who actively work against the people. Morrison writes:

> Like statist language, censored and censoring. Ruthless in its policing duties, it has no desire or purpose other than maintaining the free range of its own narcotic narcissism, its own exclusivity and dominance. However moribund, it is not without effect for it actively thwarts the intellect, stalls conscience, suppresses human potential. Unreceptive to interrogation, it cannot form or tolerate new ideas, shape other thoughts, tell another story, fill baffling silences. Official language smitheryed to sanction ignorance and preserve privilege is a suit of armor polished to shocking glitter, a husk from which the knight departed long ago. Yet there it is: dumb, predatory, sentimental. Exciting reverence in schoolchildren, providing shelter for despots, summoning false memories of stability, harmony among the public.

Yet there it is: dumb, predatory, but not so sentimental, a "yes-we-can' that means "no-we-cannot", super PACs that take us apart bit-by-bit, ALEC that will do the same drip-by-drip. Resist the deceptive self aggrandizing polemics of despots, one of whom reportedly gave $35,000,000 to try to buy the next President of the United States. Still more mind boggling is the article in the *New York Times* on June 13, 2012,[78] "Campaign Aid Is Now Surging Into 8 Figures", which states that "A group of conservative donors led by Charles and David Koch, for example, have pledged to raise as much as $400 million for issue groups, including the Koch-founded Americans for Prosperity."

Do not be taken in. Read Gene Sharp's *From Dictatorship to Democracy: A Conceptual Framework for Liberation*.[79] Organize global cafés in any place where there is room for a few people to meet. Explore other options. Participate in ethical responsible resistance. Find ways to act.

Earth Summit Rio+20
"We The Next Generation Demand Change"

At the time that I write, the global efforts of scientists are being dashed away by the Rio+20 disaster. The Earth Summit is taking place, and I have been data mining, reading every news report, blog, tweet, official and unofficial, that I can find on the Internet. Kumi Naidoo, the South African scholar and activist and Greenpeace International's Executive Director, for whom I have immense respect, sent a Twitter message on June 19, 2012 that would be read by people around the world. He wrote, "This is Rio Minus 20 which fails on equity, fails on ecology, fails on economy #rio+20 #earthsummit text longest suicide note in history". Naidoo's Tweet was picked up in *The Guardian*,[80] and remains in both forms as a warning to all humanity that we are actively participating in our own demise.

I have found that the US media coverage is less than abysmal, so I've skipped the *New York Times* and other American print and digital sources to find clues in *The Guardian*. Read, "Rio+20 summit must move world beyond 'grow now, clean up later'" by Connie Hedegaard,[81] who at the time was the EU Climate Commissioner. Then read "Rio+20: Earth summit dawns with stormier clouds than in 1992," by John Vidal,[82] *The Guardian*'s environment editor, who had also been in Rio for the '92 Earth summit. Vidal writes:

> The director of UNEP, Achim Steiner, has warned that pollution is killing millions of people a year, that ecosystem decline is increasing, that climate change is speeding up, and soil and ocean degradation is worsening. Steiner said: "If [the] trends continue … governments will preside over unprecedented levels of damage and degradation.

Earth systems are being pushed towards their biophysical limits."

Dame Barbara Stocking, Oxfam's director, said: "This is urgent. As the people with the least struggle to survive, the consumption habits of the richest are stripping the Earth of its resources. The situation is dire. We cannot go on living beyond the Earth's boundaries. The people suffering are the poorest. These are issues that will affect us all for ever."

Visit the website of the Group of 77 at the United Nations (G77), and then go right to the source. A copy of the June 2, 2012, (5:00 p.m.) draft of the U.N. main text: *The Future We Want: Our Common Vision* has been leaked to The Guardian.[83] Often early marked up copies of documents are more revealing than the finished text. A brief analysis of the language of the document quickly reveals that the US delegation is resisting the inclusion of "equity" and "common but differentiated responsibility" (CBDR), and is back tracking on the agreement that was made at the Earth Summit in Rio 1992. For people in the US this should come as no surprise given the well documented lack of concern about equity or CBDR by the political, corporate, and financial sectors in America.

The following excerpt lifts the curtain on the pro-corporate stance of the US, while at the same time it reveals the dynamic complexity of the global negotiations, and the tension between the US and other countries - especially the G77 – and also countries in the developed world:

52. We affirm that green economy in the context of sustainable development and poverty eradication should:

(a) respect each country's [national sovereignty [**over their natural resources in accordance with Principle 2 of the Rio Declaration –**

G77; U.S. delete] [and [the –G77] right to development- Liechtenstein, U.S. delete] [**the right of each country to choose its own vision, models and approaches towards sustainable development and policy space –G77; U.S. delete**], as well as its – **U.S., EU, Japan, RoK delete**] national circumstances, objectives and priorities with regard to the three dimensions of sustainable development [**, with a view to enhancing the implementation of the right to development –Liechtenstein; U.S. delete**];

In the end the whole text was purged, all life lost. David Naussbaum, WWWF-UK explained:[84]

What they did was take all the 'bracketed' issues out of the text altogether. Text gets bracketed when it's controversial – and here in Rio, it's proved to be controversial when it's been ambitious and looking to change the status quo.

And so the controversies have been addressed through compromise and capitulation in varying measures. The result is a weak text, lacking in much ambition in terms of clear actions and dates, and it doesn't measure up to the vision we have of a safe world for both people and nature.

Here is the final text for 52, which now was 58:

58. We affirm that green economy policies in the context of sustainable development and poverty eradication should:

(a) be consistent with international law;

(b) respect each country's national sovereignty over their natural resources taking into account its national circumstances, objectives, responsibilities, priorities and policy space with regard to the three dimensions of sustainable development;

On June 19, 2012, Jim Leape, the head of World Wildlife International (WWF), sent the following Twitter message from Rio: "Brazil has released new text. If this becomes final text, the past year of negotiations has been a colossal waste of time." Connie Hedegaard, EU Climate Commissioner, also wrote a Twitter message from Rio on June 19, 2012, "telling that nobody in that room adopting the text was happy. That's how weak it is. And they all knew. Disappointing"

The Tweets from Leape and Hedegaard were also picked up by *The Guardian*.

The final word in this interactive text is not written but spoken. The opening address at the Earth Summit was delivered by Brittany Trilford, who was seventeen and came from New Zealand.[85] Here is an excerpt from her speech:

> "I stand here with fire in my heart…We are all aware that time is ticking and we are quickly running out…You have seventy two hours to decide the fate of your children, my children and my children's children, and I start the clock now…We the next generation, demand change, demand action, so that we can have a future. We trust you in the next 72 hours to put our interests before all other interests and boldly do the right thing. I am here to fight for my future, that's why I am here. And I would like to end today by asking you to consider why you are here, and what you can do. Are you here to save face or are you here to save us?"

Epilogue
Put Your Trust In The People

My trust in people comes from my childhood experience of standing with the women, mothers and grandmothers, when there was an explosion at the mine. I can trace my carbon footprint to government policies of forced "transference" of coalminers and their families, of relocation and migration. As a woman my trust in people comes from working with families living in poverty, of experiencing with them catastrophes, both natural and manmade, of Hurricane Katrina, armed conflict in the Middle East, Hurricane Sandy, and Sandy Hook. In times of public emergency the human spirit is indomitable. In such moments our capacity for good counters our capacity for evil. In dangerous times, the steadfast determination of people to overcome what confronts them, their grit, as well as their caring and compassion, comes to the rescue of the people in their community, and very often to people across the world.

Never doubt that people have the capacity to overcome the adversities that confront them *if* the conditions in which they live provide them the opportunity to do so. It's my ideological stance. Not right. Not left. I believe that the decisions we make should be based on the wisdom we gain, not only from what happens to us, through our human experience, personal and shared, intellectual, emotional and spiritual, but also on the understandings we can gain from the physical and social sciences about our social, biological, and physical world.

For instance, if we know that the US is a negative statistical outlier in the developed world on every measure of health and well-being, then we should do something about it. If we know that extreme human activity is creating the conditions for a step change in the climate of the planet, then we should act. We should act before the ecological damage becomes cataclysmic, before there is nothing left for our kids to inherit except the inhospitable wastelands, the legacy of hollow men, who leave nothing behind them in the twilight of the world except the sound of our children's whimper.

I write this at a time when I have ten, perhaps fifteen, or maybe, at a stretch, twenty years left to live. I will not be here when the temperature rises and the weather patterns become so erratic and extreme that millions of people in the US and billions on the planet have to cope with life-threatening deprivations caused by weather related catastrophes, which will include public health emergencies, increasing global unrest, and armed conflict. I will not be here, but my granddaughter will.

So I hope you will forgive my impatience with the discussions currently taking place about how fast or how slow the temperature is rising. Do we have 50 years or 100 years before both the temperature and sea level rise and the planet can no longer sustain life as we know it? My response is that *it doesn't matter*. It doesn't matter. Our time is up. We have to *act* now.

One hundred years is no time at all. When I was a child and I sat on my grandmother's knee, she told me stories about her childhood in the 1880's. On my great grandmother's knee the stories she told stretched back even further, to the 1850's. My granddaughter was born in the year 2000, and when she was a little girl she sat on my knee and I told her stories about her grandmother and great grandmother's childhood in the 19th century, and we talked about her life stretching forward, possibly 100 years to the 22nd century. So I have participated in conversations that include the life experiences of my great grandmother, grandmother, and granddaughter over three centuries. One hundred years is a very short time indeed. For children born in the first decade of the 21st century, the 22nd century is now. A baby born in 2014 will be 86 at the turn of the next century. Let's do what we can now so that when her granddaughter sits on her knee, she has stories to tell of what we did in the 21st century to sustain the planet for *all* life forms – birds, mammals, insects, reptiles, fishes, mollusks, crustaceans, arachnids, and flowering and non-flowering plants—to continue living here on Earth.[86]

I have gone back and revisited the reVisioning in 2009 and 2010, the *Planet Under Pressure* Conference, and Rio+20, and I have also updated my research base by reading the November, 2013 *Intergovernmental Panel on Climate Change (IPCC)* documentation that is now available, before catching up with the *UN Climate Change Conference* held in Warsaw in November, 2013. I regret to write that I have found no reports of scientific evidence that mitigate the documentation presented in *Nineteen Clues: Great Transformations Can Be Achieved Through Collective Action.*

The November, 2013 IPCC report states that "warming in the climate system is unequivocal" and will "persist for many centuries even if emissions of CO_2 are stopped". The IPCC report also states that "Human influence on the climate system is clear. Here is the critical statement on page 15 of the *Summary for Policy Makers*:[87]

> Human influence has been detected in warming of the atmosphere and the ocean, in changes in the global water cycle, in reductions in snow and ice, in global mean sea level rise, and in changes in some climate extremes. This evidence for human influence has grown since AR4 [IPCC, 2009]. It is *extremely likely* that human influence has been the dominant cause of the observed warming since the mid-20th century.

In New York City at the time that I write on January 7, 2014, it is $5°F$, which is $1°F$ below the record set on January 7, in 1896. So much for global warming climate deniers might scoff. But yesterday, January

6, 2014, it was 55°F when I walked across Central Park. It is the extreme erraticism, the deviation from well established records of local weather patterns, and the *un*-fixed, swinging wildly, *ir*-regularity of climate change that are of increasing concern, because of the impact on ecosystems, and on flora and fauna, which has become *un*-glued.

In November 2013, the forsythia was in bloom in Central Park. It was a balmy month in which those jogging in the park dressed for exercising as if it were a warm day in May. Not only was it the warmest November on record for New York City it was the warmest November for the planet since 1880 when modern records were first kept. Here's NOAA's National Climatic Data Center:[88]

> "According to NOAA scientists, the globally averaged temperature over land and ocean surfaces for November 2013 was the highest for November since record keeping began in 1880. It also marked the 37th consecutive November and 345th consecutive month (more than 28 years) with a global temperature above the 20th century average."

The climate deniers no longer have any credibility. Climate change *is* unequivocal. I have included a reference to the Climate Change 2013 Physical Basis Working Group 1 Fact Sheet[89] because it is important that we understand just how many scientists are in agreement – lead and contributing authors, first order expert draft reviewers, second order expert and government reviewers; how many comments from governments on the final draft summary – and that 195 governments approved the *Summary for PolicyMakers* and the statement that *climate change is unequivocal and human activity is responsible.*

In November 2013, the UN Framework Convention on Climate Change (UNFCCC) took place in Warsaw, Poland.[90] It was the 19th UN Conference of Parties (COP19) which was supposed to result in a final universal climate agreement to be signed in Paris in 2015 that will replace the Kyoto Protocol. It was another Rio+20 fiasco. Kumi Naidoo might call it another suicide note.

An Editorial in *Nature Climate Change*, published online on 20 December, 2013,[91] entitled "Too Little Too Late?" quotes Nicholas Stern, who is the chair of the Grantham Research Institute on Climate Change and the Environment and the Centre for Climate Change Economics and Policy at the London School of Economics. Stern states, "although some progress has been made at this summit, the actions that have been agreed are simply inadequate when compared with the scale and urgency of the risks that the world faces from rising levels of greenhouse gases, and the dangers of irreversible impacts if there is delay".

Perhaps the only decision to come out of Warsaw that is of any significance is that a new branch of the UNFCCC is going to be established to deal with "loss and damage" from climate change, along with the existing branches that focus on "mitigation" and "adaptation". It is a significant development because it also points to the fact that there was agreement from the 198 participating countries that climate change is unequivocal. Essentially, the UNFCCC recognizes that there is an urgent need for the global community to address issues of loss that are incurred and damage that is done.

In response to the Warsaw UNFCCC, *The Guardian* published an article on November 19, 2013 with the advice of experts on building a coalition to tackle climate change,[92] and I have included a few of their suggestions before ending with my two cents worth.

"One option is to change the political narrative of climate change to a people-centered narrative," Esther Agbarakwe, founder of Nigerian Youth Climate Coalition, is quoted as stating. "Storytelling can play a big role in this."

"We must engage citizens in the solutions by helping them understand the problem," writes Kelly Rig, Executive Director, Global Call for Climate Action, Amsterdam, Netherlands.

"We should be more optimistic about collaboration and progress at a non-governmental level," writes Robert Laubacher, project director, MIT Climate CoLab, Cambridge Massachusetts.

It's a great segue.

If we are agreed in our willingness to act, the next question is what kinds of actions are we willing to participate in? Already in the US and in the global community there are conversations taking place about changing the political narrative of climate change to a people-centered narrative. Rio+20 and Warsaw might have ended with no global agreements worth the paper they were written upon, but the on-line *Global Visioning Consultation* in August, 2009 organized by the International Council for Science (ICSU) and the International Social Science Council (ISSC) and the ICSU *Visioning Open Forum* at UNESCO in Paris in 2010, followed by the March, 2012 *Planet Under Pressure* conference in London, created a global movement of Earth system scientists, civil society, and other non-governmental organizations. Amongst these efforts are those of the ISSC, "energizing social science for action" together with the Swedish International Development Agency (Sida), the UN Educational, Scientific and Cultural Organization (UNESCO), the UN Research Institute for Social Development (UNRISD), IHDP, and global regional social science councils. ISSC states:[93]

> Rapidly changing global realities drive the ever-growing demand for social science knowledge that works to inform effective and urgent responses to some of the most defining challenges of our times. In the face of ever-expanding environmental problems and disaster risks on the one hand, and converging crises of climate, inequality, food,

water, finance and social discontent on the other, the focus of demand falls sharply on innovative, sustainable social solutions.

Scientists in both the physical and social sciences are actively engaging with citizens to help them understand how dangerously near to the planetary tipping points we have come. Social media has created a public space for a ground swell of individuals and groups who are actively engaged in getting the word out that it is imperative that we alter course and take another pathway, another route, to protect the planet and our children. Many of the Earth system science sites are included at the end of *Nineteen Clues*. The Stockholm Resilience Centre is exceptionally informative, and the Potsdam Institute for Climate Impact Research is also an essential site.

Similarly, there is a groundswell on social media of educators and parents who are uniting to resist and *act* to reject the mind-manipulation which is clearly set forth in the Rice-Klein doctrine for the monetization and militarization of public education. Parenthetically, what I did not realize when I was researching the clues is that the Rice-Klein report was integral to the PR propaganda that was used to promote the developmentally inappropriate Common Core and the equally developmental inappropriate testing regimes which are chronically abusive to children *and* teachers. While there is much in the Rice – Klein report that seems rational and worth supporting, the underlying assumptions and stated goals are toxic for children and the planet.

The letter written by the New York Principals Association[94] actively engages the public in joining with educators to change the political narrative to a narrative that promotes the health and well-being of children, as well as their academic development. The New York Principals' letter not only engages parents, but also invites the public to join with them in their resistance to developmentally inappropriate testing and instruction.

"As dedicated administrators, we have carefully observed the testing process and have learned a great deal about these tests and their impact," the principals write. We care deeply about your children and their learning and want to share with you what we know – and what we do not know—about these new state assessments."

The principals provide a detailed list of their concerns, including that in New York State testing has increased dramatically, that the tests are too long, that students labeled as failures are forced out of classes, and that the achievement gap is widening.

"Children have reacted viscerally to the test," the principals state. "We know that many children cried during or after testing, and others vomited or lost control of their bowels or bladders. Others simply gave up. One teacher reported that a student kept banging his head on the desk, and wrote, 'This is too hard', and 'I can't do this,' throughout the booklet."

As of November 13, 2013, the letter had been signed by 545 New York State Principals and by nearly 3000 parents and members of the public in support.

There are comparisons that can be made between the attitude of policy makers, corporations and ideological elites, the hollow men who are hollow to the core, in their come-hell-or-high-water harnessing of the energy sources of the planet, and the come-hell-or-high-water harnessing of the skills of our children to ensure we can compete in a global economy. With their eyes on the prize of power, privilege, and profits, whatever the damage to the planet or to children, and whatever the consequences, their eyes do not see, their hearts do not feel, because for them the suffering they are causing is not real.

"In the fall of 2012," Mary Calamia, LCSW, CASAC, stated on October 10, 2013, in her testimony before the New York State Assembly,[95] "I started to receive an inordinate number of students referrals."

"A large number of honors students," Calamia continued, "mostly 8th graders, were streaming into my practice. The kids were self-mutilating, cutting themselves with sharp objects and burning themselves with cigarettes. My phone never stopped ringing."

"I also started to receive more calls referring elementary school students who were refusing to go to school," Calamia told the official state committee. "For the first time I heard the term "Common Core". Calamia spoke of "self-mutilating behaviors, insomnia, panic attacks, loss of appetite, depressed mood and in one case, suicidal thoughts that resulted in a two-week hospital stay for an adolescent."

"Are kids too coddled?" Frank Bruni asks in The *New York Times*, November 24, 2013.[96] Referring specifically to Calamia's testimony Bruni states, "If children are unraveling to this extent, it's a grave problem", but he cautions that "we need to ask ourselves how much panic is trickling down to kids from their parents and whether we're paying the price of having insulated kids from blows to their egos".

In a response that I wrote to Bruni,[97] I suggested that he read the research presented by Wilkinson and Pickett, and ask himself why it is that children in the US are a negative statistical outlier on every international comparative analysis of health, well-being, and academic development. I argued that the situation *is* very grave indeed, that we are in the middle of a national emergency, and that the people who are speaking up have been so disenfranchised that no public official or newspaper reporter is listening to them.

The immorality of human recklessness which characterizes the officially mandated abuse of our children has reached a tipping point, and a national revolt is underway harnessing social media. It is fueled by the concomitant denigration of teachers, parents, and social workers by policy makers and public officials, including the Secretary of Education Arne Duncan, and by similar attacks by mainstream news organizations, including the New York Times. It is a resistance movement that is framed by the morality of caring about children and the ways in which they are educated, and of being bold in speaking back to public officials who are instrumental in enforcing national and state policies and mandates that have the potential not only to cripple children (as if that was not enough), but also to bring down the US K-12 public education system. The scope, scale, and energy of the push-back by teachers and parents is increasing daily. Many new sites of resistance are being formed through social media across the US, and there is an international presence on many of these sites which reflects perhaps, the global rejection of the world domination of the educational markets by Pearson. A list of websites and Facebook pages is included after the Epilogue.

Teachers have always been public intellectuals, even though they have seldom been recognized as such, but in recent years they have been so denigrated, so beaten up, that it has been increasingly difficult for them to hold on to that position. The *Badass Teacher Association (BAT)* is changing that. "Badass" is a signifier of what it takes to act with moral responsibility and courage at the beginning of the 21st century, when US civil society has become vulnerable to official sanctioned immoral activity. BAT is taking a leadership role by acting ethically when policy makers, corporations, and elite ideologues, act *un*ethically. BATs transgress. To be "badassed" is to be good, if good is defined as to act in defense of children and their families. BATs dare big thoughts and resist indoctrination. They are badassed in the sense given in the Oxford Dictionary. A badass teacher is tough and uncompromising, a formidably impressive person, so amazing, she's a badass.

"I was uneasy with the name," Diane Ravitch writes,[98] "but I got over it." Ravitch, a renowned scholar, was Assistant Secretary of Education in the administration of President George H.W. Bush, and was a member of the National Assessment Governing Board, which oversees the National Assessment of Educational Progress (NAEP) from 1997 to 2004.

She has changed course and she is now the most formidable opponent of the educational reforms that are currently taking place.

"I am honored to join your group," Ravitch writes. "The best hope for the future of our society, of public education, and of the education profession is that people stand up and resist. Say 'No'. Say it loud and say it often."

"Teachers must resist," she continues, "because you care about your students, you care about your profession. You became a teacher to make a difference in the lives of children, not to take orders and obey the dictates of someone who knows nothing about your students."

"Parents must resist, to protect their children from the harm inflicted on them by high-stakes testing," Ravitch writes, "Administrators must resist ... School board members must resist ... Students must resist ..."

"Everyone who cares about the future of our democracy must resist, because public education is under attack, and public education is a foundation stone of our democracy. We must resist the phony rhetoric of "No Child Left Behind," which leaves every child behind, and we must resist the phony rhetoric of 'Race to the Top," which makes high-stakes testing the be-all and end-all of schooling. The very notion of a 'race to the top' is inconsistent with our democratic idea of equality of educational opportunity."

"You must resist, because if you do not, we will lose public education in the United States and the teaching profession will become a job, not a profession," Ravitch encourages BAT members. "What is happening today is not about 'reform' or even 'improvement,' it is about cutting costs, reducing the status of teachers, and removing from education every last shred of joy in learning."

"Be brave," Ravitch writes. "When you stand together and raise your voices, you are powerful. Thank you for counting me as one of your own. I salute you."

"We will only be able to make change if we stand together," one BAT responds, If we stand together, we are hopefully too many voices to ignore!"

"And on the name?" another teacher writes. "I'm proud to be a Badass Teacher. For too long teachers have been silent, apathetic, doormats. No more. Proudtobeabat."

"There you have it," writes another. "These are not normal times. We must leave our comfort zones, and in so doing we'll be true models for our students. If we capitulate without a fight against the forces that want to make education (and all life) a commodity, then we'll deserve whatever happens. Resist. Resist. Resist."

"Don't agonize, organize."

"When BATs and OWL's (Old Wise Ladies) fly together, watch out mosquitoes and RATS (Rhee-formers Attacking Teachers and Students)!"

"I love it! It needs to be a bumper sticker! Tweet that!"

Within two weeks of its formation, the BadAss Teachers Association had 21,000 members. At the time that I write, January 12, 2014, the membership of BAT is 36,000 and still growing.

United Opt Out (UOO) is a national movement to end corporate educational reform that has a different organizational structure to BAT. Morna McDermott, one of the founders of UOO, explained that *United Opt Out* is "recognized nationally as a significant source for information", and that the goal of UOO is "building networks and providing assistance for community action."[99] McDermott says the organization has more than

6,500 members on Facebook and that there were more than 1,000,000 hits on the UOO website just prior to the UOO "Occupy the Department of Education 2.0" protest in Washington, on April 4-7, 2013.[100]

Among the speakers were Bess Altwerger, Rick Meyer, and Steve Krashen, all of whom are eminent scholars who are working both nationally and locally with teachers and parents in defense of public education, and the in the resistance movement against the Common Core and the over-testing of children.

At Occupy the DOE, in "Hijacked: Corporate Takeover of Literacy in Two Voices", Altwerger was the Voice of Resistance and Meyer was the Voice of Reform.[101]

"There's a reading crisis," Meyer states, in his representation of corporate reformers.

"A manufactured crisis!" Altwerger shouts back, in her representation of teachers and parents.

"English only, "American" culture," Meyer insists.

"Multilingual and multicultural!" Altwerger resists.

"Commercial literacy, basal readers."

"Critical literacy, real books for readers!"

"Failing readers!"

"Lifelong readers!"

"Teacher training and regulation!"

"Teacher research and collaboration!"

"Shut it down, heed the call!"

"Keep it up, literacy for all!"

"Our schools are not broken," Steve Krashen states at Occupy the DOE. "The problem is poverty."

"Rather than spend on standards and tests, investing in protecting our children from the effects of poverty would improve school achievement."

Krashen tells the crowd. "More important, it is the right thing to do."

"This money could be spent to protect children from the effects of poverty," Krashen continues, "on expanded and improved breakfast and lunch programs, school nurses, and improved school and public libraries, especially in high-poverty areas."

On the United Opt Out website the organizers state, "Members of this site are parents, educators, students and social activists who are dedicated to the elimination of high stakes testing in public education. We use this site to collaborate, exchange ideas, support one another, share information, and initiate collective local and national actions to end the reign of fear and terror promoted by the high stakes testing agenda."

On the original United Opt Out website which was hacked, there was a list of sister organizations and a series of position statements, including one posted on April 29, 2012 entitled *"Pearson, ALEC, and the Brave New (Corporate) World: Stand Up to Pearson Now!"* The position statement has been re-posted at *Schools Matter and At The Chalk Face*.[102]

> The curtain has been pulled aside recently from the American Legislative Exchange Council (ALEC), exposing the seedy underbelly of our democracy. Organizations like ALEC circumvent the democratic process in favor of corporations. Financial resources are used to influence public officials and provide model legislation meant to easily pass through state houses of governance. Recent examples include infamous "Stand Your Ground" laws and others that seek to limit the voting rights of marginalized populations. Education reform legislation is also part of ALEC's agenda, with substantial sponsorship from corporate funds to divert the flow of valuable taxpayer dollars away from public schools.
>
> Pearson, through connections to ALEC, has become the dominant

provider of education resources and services in the K-12 and post-secondary markets.

We at **United Opt Out National** are calling on everyone **to take a stand against** Pearson.

There are many responses to the call to boycott Pearson on the UOO website.[103]

"The level of corruption is staggering but perhaps that also means that, with concerted, strategic grassroots efforts, the corporate reform system is also vulnerable to collapse," one person writes.

"Pearson," another states, "taking the lead in making our children commodities! Cramming crappy texts and horrible tests into the classroom. This is a resource grab, children and their intellect are the commodity. Pearson and McGraw Hill, I cordially uninvite you into our lives."

What is most critical in the *United Opt Out* call for a boycott of Pearson is that UOO makes transparent how embedded Pearson is in the governmental, corporate power structures of ALEC.

Remember at the *Planet Under Pressure* conference when a delegate called out "*We are not talking about the elephant in the room*"? And simultaneously, delegates when called back, "The U.S.A!"

Remember there were murmurs of agreement around the room and another delegate called out, "Americans are the most overworked miserable people on the planet". In response I wrote, "People suffer because of us", but we suffer too. We are the negative statistical outliers whose lives have become data points on Wilkinson and Pickett's research. Go back to the fifth clue:

Inequality in America is bad for the planet as well as for people, and increases our ethical responsibility to act.

It could be rewritten as:

ALEC is bad for the planet and for people, starting with the American people but also for people throughout the world.

In the US, every aspect of our lives is impacted by ALEC: education, health, employment, women's rights, children's rights, family, employment, income, debt, mobility, poverty, racial discrimination, inequities in wealth, and violent and white collar crime. Another source for primary data on *Inequality in the United States*, May 2012, is provided by Sharon Jank and Lindsey Owens at Stanford University.[104]

Whatever route you take, if you follow the trail to the roots then the extremes of inequality in US society can be found rooted in the legislative agenda of ALEC. Add the ALEC sponsored legislative rollback of the protection of fragile ecosystems, and their legislative efforts to repeal the protection of endangered species, and we are in the middle of a slow moving catastrophe. Crown it with the ALEC platform of climate change denial, and the catastrophe is no longer slow moving, but a rapidly approaching cataclysmic step change in the planetary boundaries for life on Earth.[105]

Stated succinctly, ALEC is bad for the planet.

In *United States of ALEC—A Follow-Up* Bill Moyers unpacks ALEC[106], beginning with the environmental consequences of Canadian crude. Lisa Graves of the nonprofit Center for Media and Democracy's *Alec Exposed* website appears with Moyers on the *United States of ALEC—A Follow-up*. She provides further evidence of the pro-climate change platform of ALEC, producing documents from a whistleblower on a series of seminars held for legislators who attended the 2011 ALEC Conference, which was sponsored by BP, ExxonMobil, Chevron, and Shell. One of the documents was of a seminar entitled: "Warming up to Climate Change: The Many Benefits of Increased Atmospheric CO_2."

Moyers explains that ALEC focuses on State legislatures and crafts up to 1,000 bills a year, and that ALEC has reported a twenty percent success rate in getting bills passed. Among them are K-12 public school bills, about which Mary Bottari, also of the Center for Media and Democracy spoke to Moyers.

"In the ALEC archive," Bottari said, "there's a giant stack of school choice bills and they're fat bills, too. And it's this little slice of school choice, and that little slice of school vouchers, and it's basically a long-term agenda of how to privatize public education." Bottari explains that they asked Julie Underwood at the University of Wisconsin to take a look at the documents.

"The kind of changes that ALEC is trying to impose on public education isn't really just mind reform," Underwood told Moyers, "it's actually creating a drastically different kind of education system than what we have now."

There is no doubt that the elephant in the room is ALEC. It's a matrix, funded by hundreds of corporations, right-wing think tanks and foundations, corporate law and lobbying firms, combined with members of the US Congress, including House Speaker John Boehner, and thousands of state legislators and "fellow travelers" as Moyers refers to them, "the billionaire bankrollers of the American right". Moyers names David and Charles Koch and calls them "the billionaire businessmen behind a vast industrial empire" who "are also political activists with an agenda." Moyers adds, "Their companies and foundations have been ALEC members and funders for years."

"We are suffocating in propaganda instead of gas, slowly feeling our minds go dead," T.H. White writes in *The Book of Merlyn*[107], at the beginning of the Second World War.

We are too.

It is a moment in history in years to come we will never forget, but this epilogue ends here. What happens next is not clear. It's up to all of us to do what we can do. I stop writing with "put your trust in the people" in my mind and heart, hoping that we will come through and make the planet a **child safe zone**.

Postscript

Brian Cambourne, University of Wollongong, Australia

The "Cascading" Metaphor

Denny Taylor moves - cascades - the reader across some of the Op-Ed pieces, editorials, and TV network news from 2012, the major sources of information upon which most of us rely to inform us about the dire state of affairs facing the planet and our species, and then identifies the primary sources of "hard" evidence to support (or question) the claims made in them. By providing the reader with links to these primary sources and by issuing a challenge to "data-mine" them, Taylor makes it easy for the reader to analyze the truth value of the claims. Whether they'll accept the challenge and actually "data-mine" the primary sources and then be able to devise a course of action for systemically addressing these issues is another matter. Her underlying message is that if we understand the issues and see the linkages which her ethnography identifies, then as a species we are intelligent and powerful enough to turn them around and save the world and ourselves.

The Systemic Connections Between Overlapping Domains of Concern and Inquiry

The way Taylor links global sustainability, income, and educational and gender inequality highlights and illustrates the complexity and the "connectedness" of the issues confronting us. I'm reminded of what I understand the Gaia hypothesis to be, i.e. "everything is connected to everything else." She uses her expertise as an ethnographer to move us across a range of different disciplines and domains of inquiry, subtly showing how they're linked by identifying the "clues" she identifies to help us understand the complexities of staving off the catastrophe which will overtake us if we do nothing.

The Narratives Woven Between the Identification of Each Clue

The examples of how the inequalities Taylor researched and reported in her previous books (Toxic Literacies, Growing Up Literate) are compelling. They force the reader to reflect on what she refers to as the answer to the "QoQ", i.e. "find(ing) ways to move beyond the great divisions in US society". While it's a simple phrase ("find(ing) ways to move beyond the great divisions in US society") the stories she narrates such as the struggles Laurie had and the pernicious-ness of the inequalities her life, and the current experiences of those in the '"Occupy" movement at Zuccotti Park belie the sheer simplicity of stating it as "moving beyond the great divisions in US society." Finding a way of doing it would rival the discovery of DNA as a scientific achievement.

The Embedded Model of How To Do An Ethnography

Nineteen Clues: Great Transformations Can Be Achieved Through Collective Action is a model of how ethnography should be done and used. I came away realizing that only an ethnographic perspective can capture the complexity of what's facing us. Only an ethnographic perspective can overcome the fragmentation and "silo-isation" that bedevils science today and which (in my opinion) makes concerted action on problems of this magnitude almost impossible.

Final Comment

This book needs to be published, and read by many people.

References

1. Schmidt, E. & Cohen, J. (2013). *The New Digital Age: Reshaping the Future of People, Nations and Business*. New York, NY: Alfred A. Knopf.

2. Lanier, J. (2013). *Who Owns the Future?* New York, NY: Simon & Schuster.

3. The Amsterdam Declaration on Global Change, *Challenges of a Changing Earth: Global Change Open Science Conference*, Amsterdam, The Netherlands, July 13, 2001. Retrieved from http://www.igbp.net/4.1b8ae20512db692f2a680001312.html

4. Potsdam Memorandum 2007 - A Global Contract for the Great Transformation. *Symposium on Global Sustainability: A Nobel Cause*, Potsdam, Germany, October 8-10, 2007. Retrieved from http://www.nobel-cause.de/potsdam-2007

5. International Council for Science (ICSU), *A Vision for Earth System Research: Have Your Say*, Paris, France, July 17, 2009. Retrieved from http://www.icsu.org/news-centre/press-releases/2009/a-vision-for-earth-system-research-have-your-say-1

6. International Council for Science (ICSU), *Earth System Visioning Open Forum*, Paris, France, June 22, 2010. Retrieved from http://www.globalchange.gov/whats-new/international-news/826-earth-system-visioning-open-forum.html

7. Planet Under Pressure 2012 Conference, London, England, March 26-29, 2012. *New Knowledge Towards Solutions*. Retrieved from http://www.planetunderpressure2012.net/

8. Lovejoy, T. (2012, April 5). The Greatest Challenge of Our Species. *New York Times*. Retrieved from http://www.nytimes.com/2012/04/06/opinion/the-greatest-challenge-of-our-species.html

9. Brito, L. & Stafford Smith, M. (March 29, 2012). *State of the Planet Declaration*. Planet Under Pressure 2012 Conference. London, England, March 26-29, 2012. Retrieved from http://www.planetunderpressure2012.net/

10. *Rio+20 United Nations Conference on Sustainable Development (UNCSD)*, Rio de Janeiro, Brazil, June 20-22, 2012. Retrieved from http://www.uncsd2012.org/index.html

11. United Nations (2000, September 8). *United Nations Millennium Declaration*. Retrieved from http://www.un.org/en/development/devagenda/millennium.shtml

12. United Nations Millennium Development Goals and Beyond 2015. *We Can End Poverty*. Retrieved from http://www.un.org/millenniumgoals/bkgd.shtml

13. United Nations Framework Convention on Climate Change (1997, December 11). *Kyoto Protocol*. Retrieved from http://unfccc.int/kyoto_protocol/items/2830.php

14. International Geosphere-Biosphere Programme, *Risky Business*, interview of Peter Höppe, Munich Re Head of Geo Risks Research, March, 2102. Retrieved from http://www.igbp.net/news/features/features/riskybusiness.5.705e080613685f74edb8000111.html

15. University of California Museum of Paleontology, *The Holocene Epoch*. Retrieved from http://www.ucmp.berkeley.edu/quaternary/holocene.php

16. International Geosphere-Biosphere Programme, *Anthropocene: An Epoch of Our Making*, March 7, 2012. Retrieved from http://www.igbp.net/news/features/features/anthropoceneanepochofourmaking.5.1081640c135c7c04eb480001082.html

17. Rockström, J., Steffen, W., et al (2009). Planetary Boundaries: Exploring the Safe Operating Space for Humanity. *Ecology and Society, 14*(2): 32. Retrieved from http://www.ecologyandsociety.org/vol14/iss2/art32/www.ecologyandsociety.org/vol14/iss2/art32/ES-2009-3180.pdf

18. Leemans, R. (Editor) et al (2012). Open Issue. *Current Opinion in Environmental Sustainability, 4*(1): 1- 158. Retrieved from http://www.sciencedirect.com/science/journal/18773435/4/1

19. International Geosphere-Biosphere Programme, *Global Change Magazine No. 78*. March 19, 2012. Retrieved from http://www.igbp.net/publications/globalchangemagazine/globalchangemagazine/globalchangemagazineno78.5.1081640c135c7c04eb48000371.html

20. The Equality Trust: Because More Equal Societies Work Better For Everyone. *Resources: The Spirit Level.* Retrieved from http://www.equalitytrust.org.uk/resources/spirit-level

21. Pickett, K.E. and Wilkinson, R. (2011). *The Spirit Level Age: Why Greater Equality Makes Societies Stronger.* London: Bloomsbury Publishing.

22. The Equality Trust: Because More Equal Societies Work Better For Everyone. *Resources: The Spirit Level: Education.* Retrieved from http://www.equalitytrust.org.uk/resources/spirit-level/education

23. Wilkinson, R. and Pickett, K.E. (2006). Health inequalities and the UK Presidency of the EU. *The Lancet*, 367(9517): 1126-1128. Retrieved from http://www.thelancet.com/journals/lancet/article/PIIS0140-6736(06)68489-4/fulltext doi:10.1016/S0140-6736(06)68489-4

24. Wilkinson, R.G. and Pickett, K.E. (2007). The problems of relative deprivation: Why some societies do better than others. *Social Science & Medicine*, 65(9): 1965-1978. Retrieved from http://www.sciencedirect.com/science/article/pii/S0277953607003036; doi: 10.1016/j.socscimed.2007.05.041

25. McNeil, D.G. Jr. (2010, May 17). Motherhood: Norway Tops List of the Best Places to be a Mother; Afghanistan Rates Worst. *New York Times.* Retrieved from http://www.nytimes.com/2010/05/18/health/18glob.html?_r=1&

26. Save the Children (May, 2010). *Women on the Front Lines of Health Care -State of the World's Mothers 2010.* Retrieved from http://www.savethechildren.org/site/c.8rKLIXMGIpI4E/b.8585863/k.9F31/State_of_the_Worlds_Mothers.htm#downloads

27. Save the Children (May, 2012). Nutrition in the First 1,000 Days - *State of the World's Mothers 2012.* Retrieved from http://www.savethechildrenweb.org/SOWM2012Interactive/SOWM2012_2/#/2/

28. United Nations Human Rights: Office of the High Commissioner for Human Rights (December 19, 1979). *Committee on the Elimination of Discrimination Against Women.* Retrieved from http://www.ohchr.org/en/hrbodies/cedaw/pages/cedawindex.aspx

29. United Nations Human Rights: Office of the High Commissioner for Human Rights (November 1989). *Convention on the Rights of the Child.* Retrieved from http://www.ohchr.org/EN/HRBodies/CRC/Pages/CRCIndex.aspx

30. Editorial. (2012, May 19). The Campaign Against Women. New York Times.

Retrieved from http://www.nytimes.com/2012/05/20/opinion/sunday/the-attack-on-women-is-real.html?_r=0

31. *The Fog of War: Eleven Lessons from the Life of Robert S. McNamara* (2003). Transcript retrieved from http://www.errolmorris.com/film/fow_transcript.html

32. ICSU/ISSC (2010). *Earth System Science for Global Sustainability: The Grand Challenges.* International Council for Science, Paris. Retrieved from http://www.icsu.org/publications/reports-and-reviews/grand-challenges/GrandChallenges_Oct2010.pdf

33. United States 112th Congress (2011-2012). House Bill H.R.5325. *Energy and Water Development and Related Agencies Appropriations Act, 2013.* Retrieved from http://thomas.loc.gov/cgi-bin/bdquery/z?d112:h.r.5325:

34. United States 112th Congress (2011-2012). House Bill H.R.5855. *Department of Homeland Security Appropriations Act, 2013.* Retrieved from http://thomas.loc.gov/cgi-bin/bdquery/z?d112:h.r.5855:

35. United States 112th Congress (2011-2012). Senate Bill S.3216. *Department of Homeland Security Appropriations Act, 2013.* Retrieved from http://thomas.loc.gov/cgi-bin/bdquery/z?d112:s.3216:

36. Associated Press. (2012, March 19). Panel Says Schools' Failings Could Threaten Economy and National Security. *New York Times.* Retrieved from http://www.nytimes.com/2012/03/20/education/panel-says-schools-failings-could-threaten-economy-and-national-security.html

37. Council on Foreign Relations (March, 2102). *Task Force Report No. 68: U.S. Education Reform and National Security.* http://www.cfr.org/united-states/uzsz-education-reform-national-security/p27618

38. Chivian, E. and Bernstein, A. (Editors). (2008). *Sustaining Life: How Human Health Depends on Biodiversity* (3rd Edition). Oxford University Press, USA.

39. Lenton, T.M., Held, H., et al. (2007). Tipping elements in the Earth's climate system. *Proceedings of the National Academy of Sciences of the United States of America, 105*(6), 1786-1973. Retrieved from http://www.pnas.org/content/105/6/1786.full; doi:10.1073/pnas.0705414105

40. Baudrillard, J. (1993). *The Transparency of Evil: Essays on Extreme Phenomenon,* Brooklyn, NY: Verso Books.

41. Scott, J. and DiMartino, C. (2009). Public Education Under New Management: A Typology of Educational Privatization Applied to New York City's Restructuring. *Peabody Journal of Education, 84,* 432-459. Retrieved from https://www.academia.edu/1288820/Public_Education_Under_New_Management_A_Typology_of_Educational_Privatization_Applied_to_New_York_Citys_Restructuring; doi:10.1080/01619560902973647

42. Chozick, A. (2012, May 7). Steering Murdoch in Scandal, Klein Put School Goals Aside. *New York Times.* Retrieved from http://www.nytimes.com/2012/05/08/business/media/scandal-distracts-klein-from-his-education-goals-at-news-corp.html

43. House of Commons, Culture, Media and Sport Committee, Eleventh Report of Session 2010-2012, Volume 1 (2012, May 1). *News International and Phone-hacking.* Retrieved from http://www.publications.parliament.uk/pa/cm201012/cmselect/cmcumeds/903/90302.htm

44. Phillips, A.M. (2012, May 11). E-Mails Provide Inside Look at Mayor's Charter School Battle. *New York Times*. Retrieved from http://www.nytimes.com/2012/05/12/nyregion/bloombergs-charter-school-battle-detailed-in-e-mails.html

45. Darling-Hammond, L. (2012, January 30). Why is Congress Redlining Our Schools? *The Nation*. Retrieved from http://www.thenation.com/article/165575/why-congress-redlining-our-schools#

46. Strauss, V. (2012, April 20). 'Talking pineapple' question on standardized test baffles students. *The Washington Post*. Retrieved from http://www.washingtonpost.com/blogs/answer-sheet/post/talking-pineapple-question-on-standardized-test-baffles-students/2012/04/20/gIQA8i01VT_blog.html

47. Taylor, D. (1998). *Beginning to Read and the Spin Doctors of Science*. Urbana, IL: National Council of Teachers of English.

48. Pearson. (2012). *Always Learning: Annual Report and Accounts 2012*. Retrieved from http://www.pearson.com/investors/financial-information/reports-and-results.html

49. Collins, G. (2012, June 21). How Texas Inflicts Bad Textbooks on Us. *New York Review of Books*. Retrieved from http://www.nybooks.com/articles/archives/2012/jun/21/how-texas-inflicts-bad-textbooks-on-us/

50. Kaufman, D. (2012, May 27). Land of Cheese and Rancor: How Did Wisconsin Become the Most Politically Divisive Place in America? *New York Times Magazine*. Retrieved from http://www.nytimes.com/2012/05/27/magazine/how-did-wisconsin-become-the-most-politically-divisive-place-in-america.html?_r=0

51. Editorial. (2012, June 16). Energy Etch A Sketch. *New York Times Sunday Review*. Retrieved from http://www.nytimes.com/2012/06/17/opinion/sunday/energy-etch-a-sketch.html

52. Liptak, A. (2010, January 21). Justices, 5-4, Reject Corporate Spending Limit. *New York Times*. Retrieved from http://www.nytimes.com/2010/01/22/us/politics/22scotus.html?pagewanted=all

53. Twenge, J.M., Gentile, B., et al. (2010). Birth cohort increases in psychopathology among young Americans, 1938–2007: A cross-temporal meta-analysis of the MMPI. *Clinical Psychology Review, 30*(2), 145-154. Retrieved from http://www.sciencedirect.com/science/article/pii/S027273580900141X; doi:10.1016/j.cpr.2009.10.005

54. United Nations Department of Economic and Social Affairs (2005). *The Inequality Predicament: Report on the World Social Situation 2005*. Retrieved from http://hdrnet.org/61/

55. Education Week: Editorial Projects in Education (EPE) Research Center (2011, June 9). *Diplomas Count 2011, Beyond High School, Before Baccalaureate, Meaningful Alternatives to a Four-Year Degree. Dropout Epicenters: 2011*. Retrieved from www.edweek.org/media/v30-34analysis-dropout-c4.pdf

56. Elsie (2012, May 12). Letter in response to "Why Don't We Have Any White Kids? New York Times. Retrieved from http://www.nytimes.com/2012/05/13/education/at-explore-charter-school-a-portrait-of-segregated-education.html?comments#permid=190

57. Kleinfeld, N.R. (2012, May 11). Why Don't We Have Any White Kids? *New York Times*. Retrieved from http://www.nytimes.com/2012/05/13/education/at-explore-charter-school-a-portrait-of-segregated-education.html?_r=2&&pagewanted=all

58. Editorial. (2011, October 8). Protesters Against Wall Street. *New York Times* Sunday Review. Retrieved from http://www.nytimes.com/2011/10/09/opinion/sunday/protesters-against-wall-street.html

59. Lakoff, G. (2011, October 25). How to frame yourself: A framing memo for Occupy Wall Street. *The Berkeley Blog*. Retrieved from http://blogs.berkeley.edu/2011/10/25/how-to-frame-yourself-a-framing-memo-for-occupy-wall-street/comment-page-1/

60. Bull, M. (2012, May 24). What is the rational response? *London Review of Books*, 34(10): 3-6. Retrieved from http://www.lrb.co.uk/v34/n10/malcolm-bull/what-is-the-rational-response

61. Gardiner, S. M. (2011). *A Perfect Moral Storm: The Ethical Tragedy of Climate Change*. Oxford University Press, USA.

62. Krugman, P. (2012, May 24). Egos and Immorality. *New York Times*. Retrieved from http://www.nytimes.com/2012/05/25/opinion/krugman-egos-and-immorality.html?_r=0

63. Sharp, G. (1973). *The Politics of Nonviolent Action*. Boston, MA: Porter Sargent.

64. Taylor, D. (1998).*Growing Up Literate*. Portsmouth, NH: Heinemann.

65. Taylor, D. (1996).*Toxic Literacies*. Portsmouth, NH: Heinemann.

66. Smith, T.K. (2011). *Life on Mars: Poems*. Minneapolis, MN: Graywolf Press.

67. Rogers, D.S., Duraiappah, A.K., et al. (2012). A vision for human well-being: transition to social sustainability. *Current Opinion in Environmental Sustainability*,4(1): 61-73. Retrieved from http://www.sciencedirect.com/science/article/pii/S1877343512000140; doi:10.1016/j.cosust.2012.01.013

68. United Nations General Assembly: *Report of the UN Conference on Environment and Development*, A/CONF.151/26 (Vol. I), Rio de Janeiro, Brazil, June 3-14, 1992. Retrieved from http://www.un.org/documents/ga/conf151/aconf15126-1annex1.htm

69. Hansen, J. (2012, May 9). Game Over for Climate Change. *New York Times*. Retrieved from http://www.nytimes.com/2012/05/10/opinion/game-over-for-the-climate.html

70. Yang, S. (2012, June 12). Analysis of global fire risk shows big, fast changes ahead. *University of Berkeley News Center*. Retrieved from http://newscenter.berkeley.edu/2012/06/12/climate-change-global-fire-risk/

71. Moritz, M.A., Parisien, M., et al. (2012). Climate change and disruptions to global fire activity. *Ecosphere* 3(6): Art49. Retrieved from http://www.esajournals.org/doi/abs/10.1890/ES11-00345.1; doi:10.1890/ES11-00345.1

72. Vidal, J. (2012, June 7). Many treaties to save the earth, but where's the will to implement them? *The Guardian*. Retrieved from http://www.theguardian.com/environment/blog/2012/jun/07/earth-treaties-environmental-agreements

73. United Nations Environment Programme (2012). *GEO-5 Global Environment Outlook: Environment for the Future We Want*. Retrieved from http://www.unep.org/geo/geo5.asp

74. United Nations Environment Programme (2012, January 31). *GEO-5 Global Environment Outlook: Summary for Policy Makers*. Retrieved from http://www.unep.org/geo/GEO5_SPM.asp

75. United Nations Environment Programme Press Release (2012, June 12). *World Remains on Unsustainable Track Despite Hundreds of Internationally Agreed Goals and Objectives*. Retrieved from http://www.unep.org/geo/geo5.asp

76. Watts, J. (2012, June 12). Rio+20 conference's search for green solutions hampered by deep divisions. *The Guardian*. Retrieved from http://www.theguardian.com/environment/2012/jun/12/rio-20-earth-summit-global-climate-talks

77. Morrison, T. (1993, December 7). *Nobel Lecture*. The Nobel Prize in Literature 1993. Retrieved from http://www.nobelprize.org/nobel_prizes/literature/laureates/1993/morrison-lecture.html

78. Confessore, N. (2012, June 13). Campaign Aid Is Now Surging Into 8 Figures. *New York Times*. Retrieved from http://www.nytimes.com/2012/06/14/us/politics/sheldon-adelson-sets-new-standard-as-campaign-aid-surges-into-8-figures.html?_r=1&

79. Sharp, G. (1973). *From Dictatorship to Democracy: A Conceptual Framework for Liberation*. Boston, MA: The Albert Einstein Institution.

80. Vaughan, A. (2012, June 20). Rio+20 summit: Opening day live blog. *The Guardian*. Retrieved from http://www.theguardian.com/environment/blog/2012/jun/20/rio-20-earth-summit-live-blog

81. Hedegaard, C. (2012, June 19). Rio+20 summit must move world beyond 'grow now, clean up later'. *The Guardian*. Retrieved from http://www.theguardian.com/environment/2012/jun/19/rio-20-summit-growth-sustainability

82. Vidal J. (2012, June 19). Rio+20: Earth summit dawns with stormier clouds than in 1992. *The Guardian*. Retrieved from http://www.theguardian.com/environment/2012/jun/19/rio-20-earth-summit-1992-2012

83. Rio+20 United Nations Conference on Sustainable Development (UNCSD), Rio de Janeiro, Brazil, June 20-22, 2012. Draft report (2012, June 2, 5:00 pm). *The Future We Want*. Retrieved from www.greenpeace.org/international/Global/international/publications/RioPlus20/230The_Future_We_Want-_Co-Chairs_Consolidated_Text_2_June_fo_1.pdf

84. Nussbaum, D. (2012, June 20). Will world leaders get serious at Rio? *World Wildlife Fund (WWF) UK Blog*. Retrieved from http://blogs.wwf.org.uk/blog/business-government/green-economy/will-world-leaders-get-serious-at-rio/

85. *Rio+20 United Nations Conference on Sustainable Development (UNCSD)*, Rio de Janeiro, Brazil, June 20-22, 2012. Brittany Trilford Speech to UN Rio+20 Summit Opening Ceremony. Retrieved from http://adoptanegotiator.org/2012/06/20/72-hours-to-decide-the-fate-of-your-children-my-children-my-childrens-children/

86. Opinion (2012, June 1). Are We in the Midst of a Sixth Mass Extinction? *New York Times Sunday Review*. Retrieved from http://www.nytimes.com/interactive/2012/06/01/opinion/sunday/are-we-in-the-midst-of-a-sixth-mass-extinction.html?_r=2&

87. Stocker, T.F., Qin, D. et al. (2013, September 27). *Summary for Policymakers: Climate Change 2013: The Physical Science Basis*. Contribution of Working Group I to the Fifth Assessment Report of the Intergovernmental Panel on Climate Change (IPCC). Retrieved from http://www.climatechange2013.org/

88. National Climatic Data Center (NCDC), National Oceanic and Atmospheric Administration (NOAA). (2013, November). *NCDC releases Nov 2013 Global Climate*

Report. Retrieved from http://www.ncdc.noaa.gov/news/ncdc-releases-november-2013-global-climate-report

89. Intergovernmental Panel on Climate Change (IPCC). (2013, September 27). *Climate Change 2013 Physical Basis Working Group 1 Fact Sheet*. Retrieved from http://www.climatechange2013.org/background/

90. *United Nations Framework Convention on Climate Change (UNFCCC)*. (2013, November). Retrieved from http://unfccc.int/meetings/warsaw_nov_2013/meeting/7649.php

91. Editorial. (2013, December 20). Too Little, Too Late? *Nature Climate Change* 4(1), (2014). Retrieved from http://www.nature.com/nclimate/journal/v4/n1/full/nclimate2095.html; doi:10.1038/nclimate2095

92. Young, H. (2013, November 19). 13 tips on building a coalition to tackle climate change. *The Guardian*. Retrieved from http://www.theguardian.com/global-development-professionals-network/2013/nov/19/climate-change-coalition-best-bits

93. Hackmann, H. and St. Clair, A.L. (2012, May). Transformative Cornerstones of Social Science Research for Global Change. *International Social Science Council (ISSC)*. Retrieved from http://www.worldsocialscience.org/resources/publications/

94. Fougner, S, Burris, C. et al. (2013, November 21). An Open Letter to Parents of Children throughout New York State Regarding Grade 3-8 Testing. *New York Principals Association*. Retrieved from http://www.newyorkprincipals.org/letter-to-parents-about-testing

95. Calamia, M. (2013, October 20). Common Core Can Emotionally Damage Your Child! *The Independent Sentinel*. Retrieved from http://www.independentsentinel.com/common-core-can-emotionally-damage-your-child/

96. Bruni, F. (2013, November 23). Are Kids Too Coddled? *New York Times*. Retrieved from http://www.nytimes.com/2013/11/24/opinion/sunday/bruni-are-kids-too-coddled.html?_r=0

97. Taylor, D. (2013, December 13). *"Are Kids too Coddled?" Challenging Bruni's Opinion with Scientific Data and Evidence*. Retrieved from http://garnpress.com/2013/12/garn-press-editorial-are-kids-too-coddled-challenging-brunis-opinion-with-scientific-data-and-evidence/

98. Ravitch, D. (2013, July 6). My Message to the Badass Association of Teachers. *Diane Ravitch's Blog: A site to discuss better education for all*. Retrieved from http://dianeravitch.net/2013/07/06/my-message-to-the-badass-association-of-teachers/

99. McDermott, Morna (2014, January). Quotation retrieved from original *United Opt Out* web site which was hacked in April, 2014.

100. United Opt Out. (2013, April 4-7). Occupy DOE 2.0: *The Battle for Public Schools*. Retrieved from http://unitedoptout.com/archived/official-schedule-for-occupy-doe-2-0-the-battle-for-public-schools/

101. Meyer, R. and Altwerger, B. (2013, April 6). Hijacked: Corporate Takeover of Literacy in Two Voices. *United Opt Out Occupy DOE 2.0*. Retrieved from original United Opt Out web site which was hacked in April, 2014. Available at http://webcache.googleusercontent.com/search?q=cache:qZRx5U5_--sJ:unitedoptout.com/wp-content/uploads/2013/04/Bess-Altwerger-and-Rick-Meyer-speech.

docx+&cd=1&hl=en&ct=clnk&gl=us

102. Retrieved from original *United Opt Out* web site which was hacked in April, 2014. Available at http://www.schoolsmatter.info/2012_04_01_archive.html; and http://atthechalkface.com/2012/04/30/join-the-national-boycott-against-pearson/

103. Quotations retrieved from original *United Opt Out* web site which was hacked in April, 2014.

104. Jank, S. and Owens, L. (2012, May). *New Educational Materials on Inequality*. The Stanford Center on Poverty and Inequality. Retrieved from http://www.stanford.edu/group/scspi/slides_press_release.html

105. Rockström, J. (2010, July). Let the environment guide our development. *TEDGlobal 2010*. Retrieved from http://www.ted.com/talks/johan_rockstrom_let_the_environment_guide_our_development

106. Moyers, B. (2013, July 21). United States of ALEC—A Follow-Up. *Moyers & Company*. Retrieved from http://billmoyers.com/episode/full-show-united-states-of-alec-a-follow-up/

107. White, T.H. (1987; 1958). *The Book of Merlyn: The Once and Future King (Book 2)*. New York, NY: The Berkley Publishing Group.

Declarations on Human Rights

United Nations Human Rights: Office of the High Commissioner for Human Rights. *Committee on the Elimination of Discrimination Against Women (CEDAW)*. December 18, 1979.

United Nations Human Rights: Office of the High Commissioner for Human Rights. *Convention on the Rights of the Child*. November 20, 1989.

United Nations General Assembly, Fifty-fifth Session, *United Nations Millennium Declaration*, September 8, 2000.

United Nations Department of Economic and Social Affairs (2005). *The Inequality Predicament: Report on the World Social Situation 2005*.

Declarations on Climate Change

United Nations Framework Convention on Climate Change (UNFCCC). May 9, 1992.

United Nations General Assembly: *Rio Declaration on Environment and Development.* August 12, 1992.

United Nations *Kyoto Protocol To The United Nations Framework Convention On Climate Change*, December 11, 1997.

United Nations General Assembly, Fifty-fifth Session, *United Nations Millennium Declaration*, September 18, 2000.

Amsterdam Declaration on Global Change, *Challenges of a Changing Earth: Global Change Open Science Conference*, Amsterdam, The Netherlands, July 13, 2001.

Potsdam Memorandum 2007 – A Global Contract for the Great Transformation, Executive Summary from the *Symposium on Global Sustainability: A Nobel Cause*, Potsdam, Germany, October 8-10, 2007.

International Council for Science (ICSU)/International Socila Science Council (ISSC). *Earth System Science for Global Sustainability: The Grand Challenges.* October, 2010.

State of the Planet Declaration, *Planet Under Pressure Conference*, March 26-29, 2012.

United Nations Environment Programme (UNEP). *GEO-5 Global Environment Outlook: Environment for the Future We Want.* May, 2012.

International Social Science Council (ISSC). *Transformative Cornerstones of Social Science Research for Global Change.* May, 2012.

Rio+20 – United Nations Conference on Sustainable Development: *The Future We Want*, July 27, 2012.

Intergovernmental Panel on Climate Change (IPCC). *Climate Change 2013: The Physical Science Basis.* September 27, 2013.

Climate Change: Organizations and Resources

Diversitas

Future Earth

Global Carbon Project (GCP)

Intergovernmental Panel on Climate Change (IPCC)

International Council for Science (ICSU)

International Geosphere-Biosphere Programme (IGBP)

International Human Dimensions Programme on Global Environmental Change (IHDP)

International Social Science Council (ISSC)

National Aeronautics and Space Administration (NASA)

National Oceanic and Atmospheric Administration (NOAA)

Nobel Cause Interdisciplinary Symposia

Planet Under Pressure Conference (2012)

Potsdam Institute for Climate Impact Research (PIK)

Stockholm Resilience Centre

The Royal Society

UK Met Office

United Nations Environment Program (UNEP)

United Nations Framework Convention on Climate Change (UNFCCC)

United Nations System Work on Climate Change

US National Academy of Sciences

World Climate Research Programme (WCRP)

World Meteorological Organization (WMO)

Resistance to the Corporate Education Revolution

National Organizations and Resources

At The Chalk Face

Badass Teacher Association

Change The Stakes

Children Are More than Test Scores

Citizens United for Responsible Education (CURE)

Class Size Matters

Coalition for Public Education/Coalición por la EducaciónPública

Diane Ravitch

Education Liberty Watch

Educational Alchemy

Edutopia

Fair Test - The National Center for Fair and Open Testing

Fight Common Core (American Principles Project)

Going Public

Independent Community of Educators

Keep Education Local

Lace To The Top - Students Are More Than A Test Score!

Living in Dialogue (Anthony Cody Education Week Blog)

Mercedes Schneider

No Common Sense Education

Opportunity Action – Demanding Equity and Excellence in Education

Opt Out of State Standardized Tests

Parents Across America (PAA)

Parents And Educators Against Common Core Standards

Parents & Kids Against Standardized Testing

Rethinking Schools

Save Our Schools

Say No To Common Core

Stop Common Core – Reclaiming Local Control in Education

Susan Ohanian

The Answer Sheet (Valerie Strauss - Washington Post)

The Network For Public Education

Truth About Education – The Children's Voices

http://stopcommoncore.com/Truth in American Education

United Opt Out: The Movement to End Corporate Education Reform

Yong Zhao

National Organizations and Resources - Facebook Pages

At The Chalk Face

Badass Parents Association

Badass Parents Association (Group)

Badass Teacher Association

Badass Teacher Association (Group)

Can't Be Neutral

Change The Stakes

Choose To Refuse Common Core (Group)

Citizen Action for Fair Education (CAFE) (Group)

Citizens United for Responsible Education (CURE)

Class Size Matters

Common Core Critics (Group)

Common Core Critics – National Opt Out & Refuse the Test Campaign (Group)

Dump Duncan (Group)

Fair Test - The National Center for Fair and Open Testing

Fight Common Core (American Principles Project)

Going Public

Lace To The Top (Group)

Moms Against Duncan (MAD) (Group)

Opportunity Action – Demanding Equity and Excellence in Education

Opt Out of State Standardized Tests - National (Group)

Opt Out Of The State Test: The National Movement (United Opt Out) (Group)

National Opt Out & Refuse the Test Campaign

Parents Across America (PAA)

Parents and Educators Against Common Core Standards

Parents and Educators Against Common Core Standards (Group)

Parents and Teachers Against the Common Core (Group)

Rethinking Schools

Save Our Schools

Special Ed Advocates to Stop Common Core (Group)

Stop Common Core

Principals with Principles (Group)

Teachers' Letters to Obama (Group)

The Network For Public Education

Truth in American Education

United Opt Out National

State Organizations and Resources

Alabama - Stop Common Core in Alabama

Alaska - Stop the Common Core in Alaska

Arizona - Arizonans Against Common Core

Arkansas – Arkansas Against Common Core

California - Californians United Against Common Core

California – Parents for Public Schools (San Francisco)

Colorado – Denverites for Excellent Neighborhood School Education

Connecticut - Connecticut Coalition for Social Justice in Education Funding

Florida – Florida Parents Against Common Core

Florida – Floridians Against Common Core Education

Georgia - Stop Common Core in Georgia

Idaho - Idahoans for Local Education

Illinois- Parents United for Responsible Education (Chicago)

Illinois – Stop Common Core Illinois

Indiana – Hoosiers Against Common Core

Iowa - Iowans for Local Control

Kansas - Kansans Against Common Core

Kentucky - Kentuckians Against Common Core Standards

Maine - No Common Core Maine

Massachusetts - Can't Be Neutral

Michigan - Stop Common Core in Michigan

Minnesota - Minnesotans Against Common Core

Missouri – Missouri Coalition Against Common Core

Missouri – Missouri Education Watchdog

Nevada – Stop Common Core in Nevada

New Jersey - Education Law Center

New York - Abolish Common Core

New York – Allies for Public Education

New York - Campaign for Fiscal Equity

New York – Change The Stakes

New York - Children should not be a number (NYS Stop Testing)

New York - Coalition for Public Education (NYC)

New York - Concerned Advocates for Public Education (NYC)

New York - Ed Notes Online (NYC)

New York – Education New York

New York - Grassroots Education Movement

New York - Independent Commission on Public Education (NYC)

New York – New York Collective of Radical Educators

New York – New York Principals

New York – No Common Sense Education

New York - NYC Educator

New York - NYC Public School Parents (NYC)

New York - NYS Allies For Public Education

New York - Parent Voices NY

New York - Stop Common Core in New York State

New York - Stop Common Core in NY

New York - Teachers Unite (NYC)

New York - WNYers for Public Education

North Carolina – Durham Allies for Responsive Education (Durham)

Ohio - Ohioans Against Common Core

Oklahoma - Restore Oklahoma Public Education (ROPE)

Oregon – Oregon Save Our Schools

Oregon - Stop Common Core in Oregon

Pennsylvanian - Pennsylvanians Against Common Core

Pennsylvania – Parents United for Public Education (Philadelphia)

South Carolina – South Carolina Parents Involved in Education (SCPIE)

South Dakota - South Dakotans Against Common Core

Tennessee – Tennessee Against Common Core

Utah – Common Core: Education Without Representation

Utah - Utahns Against Common Core

Washington - Stop Common Core in Washington State

Wyoming – Wyoming Freedom in Education

Wyoming – Wyoming Against the Common Core

State Organization and Resources on Facebook

Alabama - Stop Common Core In Alabama

Alabama - Alabamians Against Common Core Standards in Education

Alaska - Alaskans Against the Common Core

Arizona - Mohave County Against Common Core (Group)

Arizona - Stop Common Core in Arizona (Group)

Arkansas - Arkansas Against Common Core (Group)

Arkansas - Arkansas Against Common Core

California - Californians Against Common Core (Group)

California - Stop Common Core in California

Colorado - Colorado Against Common Core

Colorado - Mesa County Citizens/Businesses Against Common Core Curriculum

Colorado - Parents and Educators Against Common Core Curriculum in Colorado

Colorado - Parent LED reform

Colorado - Stop Common Core in Colorado

Connecticut - Coalition for Social Justice in Education Funding

Connecticut - Stop Common Core In CT

Delaware - Delaware Against Common Core (Group)

Delaware - Delaware Against Common Core

Florida – Florida Common Core Watch

Florida - Stop Common Core in Florida

Florida – Central Florida Parents Against Common Core (FPACC) (Group)

Georgia - Stop Common Core In Georgia

Georgia - Georgians to Stop Common Core (Group)

Hawaii – Stop Common Core in Hawaii

Idaho - Idahoans Against Common Core

Idaho - Idahoans for Local Education

Idaho – Stop Common Core in Idaho

Illinois - Stop Common Core in Illinois

Indiana – Hoosiers Against Common Core

Indiana - Hoosier Moms Say No To Common Core

Iowa - Iowans for Local Control

Iowa – Stop Common Core in Iowa

Kansas - Kansans Against Common Core

Kentucky - Kentuckians Against Common Core Standards

Louisiana – Concerned Parents Against Common Core (Group)

Louisiana - Parents and Educators Against Common Core in Louisiana

Louisiana - Stop Common Core in Louisiana

Maine - No Common Core Maine

Maryland - Marylanders Against Common Core

Maryland - Stop Common Core in Maryland

Maryland - Stop Common Core in Maryland (Group)

Massachusetts - Can't Be Neutral

Michigan - Stop Common Core in Michigan

Minnesota – Minnesota Against Common Core

Mississippi - Stop Common Core In Mississippi

Missouri – Missouri for the Removal of Common Core

Missouri – Missouri Education Watchdog

Missouri – Missouri For Local Education (Group)

Montana – Montana Against Common Core (Group)

Montana - Stop Common Core in Montana

Nebraska - Nebraska Parents & Educators Against Common Core

Nevada - Nevadans Against Common Core (Group)

Nevada - Nevada Parents & Teachers STOP Common Core (Group)

Nevada - Nevada Parents & Teachers STOP Common Core

Nevada - Parent Led Reform Nevada - Stop Common Core (Group)

New Hampshire - School Choice for New Hampshire

New Hampshire - Stop Common Core in New Hampshire

New Hampshire - NH Families for Education (Group)

New Jersey - Cure NJ

New Jersey - Citizen Action for Fair Education (CAFE) (Group)

New Jersey - The Committee to Combat Common Core Curriculum (Group)

New Mexico - Stop Common Core In New Mexico

New York - New York Grassroots Against Common Core (Group)

New York - Heads Down, Thumbs Up, Hudson Valley, N.Y.

New York - Long Island Opt-Out Info (Group)

New York - Long Island Parents and Teachers Against Standardized Testing & APPR

New York - Long Islanders United Against the Common Core

New York - NY Parents Opposed to Data Sharing without Consent! (Group)

New York - New York BATs (Group)

New York - Opt Out of State Standardized Tests (Group)

New York - Oswego County for Public Education Discussion Group (Group)

New York - Parents & Teachers Against Common Core (Group)

New York - Pencils DOWN Rockland County

New York - NYS Refuse the Tests (Group)

New York - Rethinking Testing: Mid-Hudson Region

New York - STACCT: Southern Tier Against Common Core and Testing (Group)

New York - Staten Island - Know "Common Core" (Group)

New York - Stop Common Core in Cayuga/Onondaga County (Group)

New York - Stop Common Core in New York Catholic Schools (Group)

New York - Stop Common Core in New York -Monroe County (Group)

New York – Stop Common Core in NY

New York - Stop Common Core in New York State (Group)

New York - Stop Common Core in NYS Dutchess County (Group)

New York - Stop Common Core in Westchester County, NY (Group)

New York - Video Blast! Tell your New York State Common Core Story (Group)

North Carolina - Stop Common Core in NC

North Dakota - Stop Common Core in North Dakota

Ohio – Boardman Spartans Against Common Core (Group)

Ohio Educators and Parents Against Common Core Curriculum

Ohio - Ohio Common Core - Reality of Education Standards & Reform

Ohio - Ohio Parents and Teachers Against Common Core (Group)

Ohio - Stop Common Core in Ohio (Group)

Oklahoma - Oklahoma Parents and Educators for Public Education (Group)

Oklahoma - Restore Oklahoma Public Education (ROPE)

Oregon - Parent Led Reform - Oregon

Oregon - Stop Common Core in Oregon (Group)

Pennsylvanian - Pennsylvanians Against Common Core

Rhode Island - Stop Common Core in Rhode Island

Rhode Island - Rhode Islanders Against Common Core

South Carolina - Stop Common Core in South Carolina

South Dakota - South Dakotans Against Common Core

Tennessee - Stop Common Core in Tennessee

Tennessee – Education Matters Institute

Texas - Texans Against CSCOPE (Group)

Utah - Utahns Against Common Core

Virginia - Against Common Core in Virginia (Group)

Virginia - Eye on Virginia Education

Virginia - Stop Common Core in Virginia (Group)

Virginia - Virginians Concerned about K12 Education (Group)

Washington - Washington State Against Common Core State Standards (Group)

Washington - Washington State Against Common Core State Standards

West Virginia - West Virginia Against Common Core

Wisconsin - Stop Common Core in Wisconsin

Wyoming - Stop Common Core in Wyoming (Group)

Other Organizations and Resources

American Principles Project - Education

Cato Institute - Common Core: The Great Debate

Common Core - Education Without Representation

Common Core Issues

Common Core Issues - Video Website

Common Core Movie Trailer for "Building the Machine"

Common Crud Scans/Photos of Common Core Curriculum Assignments) (Group)

Common Dreams

Educational Alchemy

Inappropriate Common Core Lessons (Group)

Only a Teacher (PBS Series)

Stop the National Common Core Power Grab: Reclaim Local Control of Education Video

The Educational Freedom Coalition (Educational Provider Data Base/Alignment with CCSS)

Truth in American Education (Audio/Video Resources)

Truthout

About The Author

Denny Taylor is Professor Emerita of Literacy Studies at Hofstra University, and the founder and CEO of Garn Press. She lives in a studio the size of a bookshelf in New York City. She writes every day and spends a lot of time trying not to buy any more books. Fortunately she is not very successful at not buying! Books are her greatest pleasure, and publishing them has become as exciting as writing them. She views the construction of eBooks as a craft as well as a technical feat.

Taylor is a lifelong activist and scholar committed to nurturing the imagination and human spirit, and she regards art, literature, and science as inseparable. She prefers actionable knowledge over right or left ideology, and she always wants to see the raw data rather than read the spin about it.

Taylor's doctoral dissertation was published in 1983 as *Family Literacy*, is still in print, and is regarded a classic within the field; *Growing Up Literate* received the Mina P. Shaughnessy award in 1988 from the Modern Language Association of America; and *Toxic Literacies* was nominated for both the Pulitzer Prize and the National Book Award. In 2004, Taylor was inducted into the International Reading Association's Reading Hall of Fame.

Most recently, Taylor's trans-disciplinary research has incorporated both the physical and social sciences. She participated in the June, 2010 International Science Council (ICSU) forum at UNESCO in Paris, which focused on developing a vision of new institutional frameworks for Global Sustainability Research (Earth System Visioning). She subsequently authored four peer reviewed research papers which combined data from the social and physical sciences that were presented at the *Planet Under Pressure* Conference in London in March, 2012. Taylor brings all of her experience to Garn. Caring deeply about People and the Planet, about Language and Social Policy, and about Imagination and the Human Spirit, she has made these the three imprint pillars of Garn Press.

"We are making Garn the people's press," Taylor says. "Our mission is actionable knowledge, and we want to make Garn Press synonymous with social action."

Books by Denny Taylor

Forthcoming Garn Press Books by Denny Taylor

Keys To The Future: A Teacher's Guide To Making The Earth A Child Safe Zone

Rosie's Umbrella (a novel)

Educated or Indoctrinated?

Fukushima - Can Science Save Us? Can Policy Makers Can Pass The Car Battery Test?

Death of Childhood

People And The Planet: The Great Acceleration From Adaptation to Transformation

Other Books by Denny Taylor

Beginning to Read and the Spin Doctors of Science (1998)

Many Families, Many Literacies: An International Declaration of Principles (1997)

Teaching and Advocacy (1997)

Toxic Literacies: Exposing the Injustice of Bureaucratic Texts (1996)

From The Child's Point Of View (1993)

Learning Denied: Inappropriate Educational Decision Making (1990)

Growing Up Literate, Learning From Inner City Families (1988)

Family Storybook Reading (1986)

Family Literacy: Young Children Learning to Read and Write (Second Edition, 1998)

Family Literacy: Young Children Learning to Read and Write (First Edition, 1983)

GARN PRESS

NEW YORK, NY